Erwin Dee Kord (Ed.)

Self Drying Concrete Technology (Flooring)

Erwin Dee Kord (Ed.)

Self Drying Concrete Technology (Flooring)

Portland Cement

Solv

Contents

Self Drying Concrete Technology (Flooring)

Self-drying concrete technology is found in certain cementitious patching and leveling materials and tile-setting mortars used in the flooring industry. Self-drying techology allows the cement mix to consume all of its mix water while curing, eliminating the need for excess water to evaporate prior to installing flooring. Traditional floor coverings, such as VCT, sheet vinyl, carpet and ceramic tile, can be installed before the material is completely dry and as soon as it hardens, which typically happens in the first two hours after placement.

Traditional concrete has a water:cement ratio of about 0.5, which refers to the weight of the water divided by the weight of the cement. A water:cement ratio of 0.5 provides good workability while keeping the amount of excess water in the mix fairly low. Without at least this much extra water, the concrete would be too dry to place.

The chemical reaction of Portland cement and water that is known as hydration, which is necessary for the strengthening of the concrete, requires a water:cement ratio of only about 0.25. With a water:cement ratio of 0.5, there is twice the amount of water in the concrete mix than what is needed for hydration. This excess water needs to evaporate before flooring can be installed, which typically takes 28 days. Conversely, a self-drying concrete blend consumes all of its mix water with a water:cement ratio of up to 0.6, maintaining good workability while allowing flooring to be installed before it is completely dry.

There are also cement products that are partially self-drying, meaning that they use a high percentage of their mix water for hydration as opposed to using 100% of it. This type of product might be used when the flooring does not need to be installed the same day but must still be installed more quickly than traditional concrete would allow. For instance, products that are 80% self-drying allow flooring to be installed the next day, typically after a 16-hour cure.

Self-drying technology was developed by Ardex in Germany and was introduced in the United States in 1978.

Concrete

Concrete is a composite construction material, composed of cement (commonly Portland cement) and other cementitious materials such as fly ash and slag cement, aggregate (generally a coarse aggregate made of gravel or crushed rocks such as limestone, or granite, plus a fine aggregate such as sand), water and chemical admixtures.

The word concrete comes from the Latin word "concretus" (meaning compact or condensed), the perfect passive participle of "concrescere", from "con-" (together) and "crescere" (to grow).

Concrete solidifies and hardens after mixing with water and placement due to a chemical process known as hydration. The water reacts with the cement, which bonds the other components together, eventually

Outer view of the Roman Pantheon, still the largest unreinforced solid concrete dome.[1]

creating a robust stone-like material. Concrete is used to make pavements, pipe, architectural structures, foundations, motorways/roads, bridges/overpasses, parking structures, brick/block walls, footings for gates, fences and poles and even boats.

Concrete is used more than any other man-made material in the world.[2] As of 2006, about 7.5 billion cubic meters of concrete are made each year—more than one cubic meter for every person on Earth.[3]

Concrete powers a US$35 billion industry, employing more than two million workers in the United States alone. More than 55000 miles (89000 km) of highways in the United States are paved with this material. Reinforced concrete, prestressed concrete and precast concrete are the most widely used types of concrete functional extensions in modern days.

History

Concrete was used for construction in many ancient structures.[4]

During the Roman Empire, Roman concrete (or *opus caementicium*) was made from quicklime, pozzolana and an aggregate of pumice. Its widespread use in many Roman structures, a key event in the history of architecture termed the Roman Architectural Revolution, freed Roman construction from the restrictions of stone and brick material and allowed for revolutionary new designs in terms of both structural complexity and dimension.[5]

A modern building: Boston City Hall (completed 1968) is constructed largely of concrete, both precast and poured in place.

Opus caementicium lying bare on a tomb near Rome. In contrast to modern concrete structures, the concrete walls of Roman buildings were covered, usually with brick or stone.

Concrete, as the Romans knew it, was a new and revolutionary material. Laid in the shape of arches, vaults and domes, it quickly hardened into a rigid mass, free from many of the internal thrusts and strains that troubled the builders of similar structures in stone or brick.[6]

Modern tests show that *opus caementicium* had as much compressive strength as modern Portland-cement concrete (ca. 200 kg/cm^2).[7] However, due to the absence of steel reinforcement, its tensile strength was far lower and its mode of application was also different:

Hadrian's Pantheon in Rome is an example of Roman concrete construction.

> Modern structural concrete differs from Roman concrete in two important details. First, its mix consistency is fluid and homogeneous, allowing it to be poured into forms rather than requiring hand-layering together with the placement of aggregate, which, in Roman practice, often consisted of rubble. Second, integral reinforcing steel gives modern concrete assemblies great strength in tension, whereas Roman concrete could depend only upon the strength of the concrete bonding to resist tension.[8]

The widespread use of concrete in many Roman structures has ensured that many survive to the present day. The Baths of Caracalla in Rome are just one example. Many Roman aqueducts and bridges have masonry cladding on a concrete core, as does the dome of the Pantheon.

Some have stated that the secret of concrete was lost for 13 centuries until 1756, when the British engineer John Smeaton pioneered the use of hydraulic lime in concrete, using pebbles and powdered brick as aggregate. However, the Canal du Midi was built using concrete in 1670.[9] Likewise there are concrete structures in Finland that date back to the 16th century. Portland cement was first used in concrete in the early 1840s.

Additives

Concrete additives have been used since Roman and Egyptian times, when it was discovered that adding volcanic ash to the mix allowed it to set under water. Similarly, the Romans knew that adding horse hair made concrete less liable to crack while it hardened and adding blood made it more frost-resistant.[10]

Recently the use of recycled materials as concrete ingredients has been gaining popularity because of increasingly stringent environmental legislation. The most conspicuous of these is fly ash, a by-product of coal-fired power plants. This use reduces the amount of quarrying and landfill space required as the ash acts as a cement replacement thus reducing the amount of cement required.

In modern times, researchers have experimented with the addition of other materials to create concrete with improved properties, such as higher strength or electrical conductivity. Marconite is one example.

Composition

There are many types of concrete available, created by varying the proportions of the main ingredients below. In this way or by substitution for the cemetitious and aggregate phases, the finished product can be tailored to its application with varying strength, density, or chemical and thermal resistance properties.

The *mix design* depends on the type of structure being built, how the concrete will be mixed and delivered and how it will be placed to form this structure.

Cement

Portland cement is the most common type of cement in general usage. It is a basic ingredient of concrete, mortar and plaster. English masonry worker Joseph Aspdin patented Portland cement in 1824; it was named because of its similarity in color to Portland limestone, quarried from the English Isle of Portland and used extensively in London architecture. It consists of a mixture of oxides of calcium, silicon and aluminium. Portland cement and similar materials are made by heating limestone (a source of calcium) with clay and grinding this product (called *clinker*) with a source of sulfate (most commonly gypsum).

Water

Combining water with a cementitious material forms a cement paste by the process of hydration. The cement paste glues the aggregate together, fills voids within it and allows it to flow more freely.

Less water in the cement paste will yield a stronger, more durable concrete; more water will give a freer-flowing concrete with a higher slump. Impure water used to make concrete can cause problems when setting or in causing premature failure of the structure.

Hydration involves many different reactions, often occurring at the same time. As the reactions proceed, the products of the cement hydration process gradually bond together the individual sand and gravel particles and other components of the concrete, to form a solid mass.

Reaction:

Cement chemist notation: $C_3S + H \rightarrow C\text{-}S\text{-}H + CH$

Standard notation: $Ca_3SiO_5 + H_2O \rightarrow (CaO) \cdot (SiO_2) \cdot (H_2O)(gel) + Ca(OH)_2$

Balanced: $2Ca_3SiO_5 + 7H_2O \rightarrow 3(CaO) \cdot 2(SiO_2) \cdot 4(H_2O)(gel) + 3Ca(OH)_2$

Aggregates

Fine and coarse aggregates make up the bulk of a concrete mixture. Sand, natural gravel and crushed stone are used mainly for this purpose. Recycled aggregates (from construction, demolition and excavation waste) are increasingly used as partial replacements of natural aggregates, while a number of manufactured aggregates, including air-cooled blast furnace slag and bottom ash are also permitted.

Decorative stones such as quartzite, small river stones or crushed glass are sometimes added to the surface of concrete for a decorative "exposed aggregate" finish, popular among landscape designers.

The presence of aggregate greatly increases the robustness of concrete above that of cement, which otherwise is a brittle material and thus concrete is a true composite material.

Redistribution of aggregates after compaction often creates inhomogeneity due to the influence of vibration. This can lead to strength gradients.[11]

Reinforcement

Concrete is strong in compression, as the aggregate efficiently carries the compression load. However, it is weak in tension as the cement holding the aggregate in place can crack, allowing the structure to fail. Reinforced concrete solves these problems by adding either steel reinforcing bars, steel fibers, glass fiber, or plastic fiber to carry tensile loads. Thereafter the concrete is reinforced to withstand the tensile loads upon it.

Installing rebar in a floor slab during a concrete pour.

Chemical admixtures

Chemical admixtures are materials in the form of powder or fluids that are added to the concrete to give it certain characteristics not obtainable with plain concrete mixes. In normal use, admixture dosages are less than 5% by mass of cement and are added to the concrete at the time of batching/mixing.[12] The common types of admixtures[13] are as follows.

- Accelerators speed up the hydration (hardening) of the concrete. Typical materials used are $CaCl_2$, $Ca(NO_3)_2$ and $NaNO_3$. However, use of chlorides may cause corrosion in steel reinforcing and is prohibited in some countries, so that nitrates may be favored.
- Retarders slow the hydration of concrete and are used in large or difficult pours where partial setting before the pour is complete is undesirable. Typical polyol retarders are sugar, sucrose, sodium gluconate, glucose, citric acid, and tartaric acid.
- Air entrainments add and entrain tiny air bubbles in the concrete, which will reduce damage during freeze-thaw cycles, thereby increasing the concrete's durability. However, entrained air entails a trade off with strength, as each 1% of air may result in 5% decrease in compressive strength.
- Plasticizers increase the workability of plastic or "fresh" concrete, allowing it be placed more easily, with less consolidating effort. A typical plasticizer is lignosulfonate. Plasticizers can be used to reduce the water content of a concrete while maintaining workability and are sometimes called *water-reducers* due to this use. Such treatment improves its strength and durability characteristics. Superplasticizers (also called *high-range water-reducers*) are a class of plasticizers that have fewer deleterious effects and can be used to increase workability more than is practical with traditional plasticizers. Compounds used as superplasticizers include sulfonated naphthalene formaldehyde condensate, sulfonated melamine formaldehyde condensate, acetone formaldehyde condensate and polycarboxylate ethers.
- Pigments can be used to change the color of concrete, for aesthetics.
- Corrosion inhibitors are used to minimize the corrosion of steel and steel bars in concrete.

- Bonding agents are used to create a bond between old and new concrete (typically a type of polymer) .
- Pumping aids improve pumpability, thicken the paste and reduce separation and bleeding.

Mineral admixtures and blended cements

There are inorganic materials that also have pozzolanic or latent hydraulic properties. These very fine-grained materials are added to the concrete mix to improve the properties of concrete (mineral admixtures),[12] or as a replacement for Portland cement (blended cements).[14]

Blocks of concrete in Belo Horizonte, Brazil.

- Fly ash: A by-product of coal-fired electric generating plants, it is used to partially replace Portland cement (by up to 60% by mass). The properties of fly ash depend on the type of coal burnt. In general, siliceous fly ash is pozzolanic, while calcareous fly ash has latent hydraulic properties.[15]
- Ground granulated blast furnace slag (GGBFS or GGBS): A by-product of steel production is used to partially replace Portland cement (by up to 80% by mass). It has latent hydraulic properties.[16]
- Silica fume: A by-product of the production of silicon and ferrosilicon alloys. Silica fume is similar to fly ash, but has a particle size 100 times smaller. This results in a higher surface to volume ratio and a much faster pozzolanic reaction. Silica fume is used to increase strength and durability of concrete, but generally requires the use of superplasticizers for workability.[17]
- High reactivity Metakaolin (HRM): Metakaolin produces concrete with strength and durability similar to concrete made with silica fume. While silica fume is usually dark gray or black in color, high-reactivity metakaolin is usually bright white in color, making it the preferred choice for architectural concrete where appearance is important.

Concrete production

The processes used vary dramatically, from hand tools to heavy industry, but result in the concrete being placed where it cures into a final form. Wide range of technological factors may occur during production of concrete elements and their influence to basic characteristics may vary.[18]

When initially mixed together, Portland cement and water rapidly form a gel, formed of tangled chains of interlocking crystals. These continue to react over time, with the initially fluid gel often aiding in placement by improving workability. As the concrete sets, the chains of crystals join and form a rigid structure, gluing the aggregate particles in place. During curing, more of the cement reacts with the residual water (hydration).

Concrete plant facility (background) with concrete delivery trucks.

This curing process develops physical and chemical properties. Among these qualities are mechanical strength, low moisture permeability and chemical and volumetric stability.

Mixing concrete

Thorough mixing is essential for the production of uniform, high quality concrete. For this reason equipment and methods should be capable of effectively mixing concrete materials containing the largest specified aggregate to produce *uniform mixtures* of the lowest slump practical for the work.

Separate paste mixing has shown that the mixing of cement and water into a paste before combining these materials with aggregates can increase the compressive strength of the resulting concrete.[19] The paste is generally mixed in a *high-speed*, shear-type mixer at a w/cm (water to cement ratio) of 0.30 to 0.45 by mass. The cement paste premix may include admixtures such as accelerators or retarders, superplasticizers, pigments, or silica fume. The premixed paste is then blended with aggregates and any remaining batch water and final mixing is completed in conventional concrete mixing equipment.[20]

High-energy mixed (HEM) concrete is produced by means of high-speed mixing of cement, water and sand with net specific energy consumption of at least 5 kilojoules per kilogram of the mix. A plasticizer or a superplasticizer is then added to the activated mixture, which can later be mixed with aggregates in a conventional concrete mixer. In this process, sand provides dissipation of energy and creates high-shear conditions on the surface of cement particles. This results in the full volume of water interacting with cement. The liquid activated mixture can be used by itself or foamed (expanded) for lightweight concrete.[21] HEM concrete hardens in low and subzero temperature conditions and possesses an increased volume of gel, which drastically reduces capillarity in solid and porous materials.

Workability

Workability is the ability of a fresh (plastic) concrete mix to fill the form/mold properly with the desired work (vibration) and without reducing the concrete's quality. Workability depends on water content, aggregate (shape and size distribution), cementitious content and age (level of hydration) and can be modified by adding chemical admixtures, like superplasticizer. Raising the water content or adding chemical admixtures will increase concrete workability. Excessive water will lead to increased bleeding (surface water) and/or segregation of aggregates (when the cement and aggregates start to separate), with the resulting concrete having reduced quality. The use of an aggregate with an undesirable gradation can result in a very harsh mix design with a very low slump, which cannot be readily made more workable by addition of reasonable amounts of water.

Pouring and smoothing out concrete at Palisades Park in Washington DC.

Workability can be measured by the concrete slump test, a simplistic measure of the plasticity of a fresh batch of concrete following the ASTM C 143 or EN 12350-2 test standards. Slump is normally measured by filling an "Abrams cone" with a sample from a fresh batch of concrete. The cone is placed with the wide end down onto a level, non-absorptive surface. It is then filled in three layers of equal volume, with each layer being tamped with a steel rod in order to consolidate the layer. When the cone is carefully lifted off, the enclosed material will slump a certain amount due to gravity. A relatively dry sample will slump very little, having a slump value of one or two inches (25 or 50 mm). A relatively wet concrete sample may slump as much as eight inches. Workability can also be measured by using the flow table test.

Slump can be increased by addition of chemical admixtures such as plasticizer or superplasticizer without changing the water-cement ratio. Some other admixtures, especially air-entraining admixture, can increase the slump of a mix.

High-flow concrete, like self-consolidating concrete, is tested by other flow-measuring methods. One of these methods includes placing the cone on the narrow end and observing how the mix flows through the cone while it is gradually lifted.

After mixing, concrete is a fluid and can be pumped to the location where needed.

Curing

In all but the least critical applications, care needs to be taken to properly *cure* concrete, to achieve best strength and hardness. This happens after the concrete has been placed. Cement requires a moist, controlled environment to gain strength and harden fully. The cement paste hardens over time, initially setting and becoming rigid though very weak and gaining in strength in the weeks following. In around 3 weeks, typically over 90% of the final strength is reached, though strengthening may continue for decades.[22] The conversion of calcium hydroxide in the concrete into calcium carbonate from absorption of CO_2 over several decades further strengthen the concrete and making it more resilient to damage. However, this reaction, called carbonation, lowers the pH of the cement pore solution and can cause the reinforcement bars to corrode.

A concrete slab ponded while curing.

Hydration and hardening of concrete during the first three days is critical. Abnormally fast drying and shrinkage due to factors such as evaporation from wind during placement may lead to increased tensile stresses at a time when it has not yet gained sufficient strength, resulting in greater shrinkage cracking. The early strength of the concrete can be increased if it is kept damp during the curing process. Minimizing stress prior to curing minimizes cracking. High-early-strength concrete is designed to hydrate faster, often by increased use of cement that increases shrinkage and cracking. Strength of concrete changes (increases) up to three years. It depends on cross-section dimension of elements and conditions of structure exploitation.[23]

During this period concrete needs to be kept under controlled temperature and humid atmosphere. In practice, this is achieved by spraying or ponding the concrete surface with water, thereby protecting the concrete mass from ill effects of ambient conditions. The pictures to the right show two of many ways to achieve this, ponding – submerging setting concrete in water and wrapping in plastic to contain the water in the mix. Additional common curing methods include wet burlap and/or plastic sheeting covering the fresh concrete, or by spraying on a water-impermeable temporary curing membrane.

Properly curing concrete leads to increased strength and lower permeability and avoids cracking where the surface dries out prematurely. Care must also be taken to avoid freezing, or overheating due to the exothermic setting of cement. Improper curing can cause scaling, reduced strength, poor abrasion resistance and cracking.

Properties

Concrete has relatively high compressive strength, but much lower tensile strength. For this reason is usually reinforced with materials that are strong in tension (often steel). The elasticity of concrete is relatively constant at low stress levels but starts decreasing at higher stress levels as matrix cracking develops. Concrete has a very low coefficient of thermal expansion and shrinks as it matures. All concrete structures will crack to some extent, due to shrinkage and tension. Concrete that is subjected to long-duration forces is prone to creep.

Tests can be made to ensure the properties of concrete correspond to specifications for the application.

Environmental and health

For the environmental impact of cement production see Cement

Carbon dioxide emissions and climate change

The cement industry is one of two primary producers of carbon dioxide (CO_2), creating up to 5% of worldwide man-made emissions of this gas, of which 50% is from the chemical process and 40% from burning fuel.[24] The carbon dioxide (CO_2) produced for the manufacture of one tonne of structural concrete (using ~14% cement) is estimated at 410 kg/m^3 (~180 kg/tonne @ density of 2.3g/cm^3) (reduced to 290 kg/m^3 with 30% fly ash replacement of cement).[25] The CO_2 emission from the concrete production is directly proportional to the cement content used in the concrete mix; 900 kg of CO_2 are emitted for the fabrication of every ton of cement.[26] Cement manufacture contributes greenhouse gases both directly through the production of carbon dioxide when calcium carbonate is thermally decomposed, producing lime and carbon dioxide,[27] and also through the use of energy, particularly from the combustion of fossil fuels.

Surface runoff

Surface runoff, when water runs off impervious surfaces, such as non-porous concrete, can cause heavy soil erosion and flooding. Urban runoff tends to pick up gasoline, motor oil, heavy metals, trash and other pollutants from sidewalks, roadways and parking lots.[28] [29] Without attenuation, the impervious cover in a typical urban area limits groundwater percolation and causes five times the amount of runoff generated by a typical woodland of the same size.[30] A 2008 report by the United States National Research Council identified urban runoff as a leading source of water quality problems.[31]

Urban heat

Both concrete and asphalt are the primary contributors to what is known as the urban heat island effect.

Using light-colored concrete has proven effective in reflecting up to 50% more light than asphalt and reducing ambient temperature.[32] A low albedo value, characteristic of black asphalt, absorbs a large percentage of solar heat and contributes to the warming of cities. By paving with light colored concrete, in addition to replacing asphalt with light-colored concrete, communities can lower their average temperature.[33]

In many U.S. cities, pavement covers about 30–40% of the surface area.[32] This directly affects the temperature of the city and contributes to the urban heat island effect. Paving with light-colored concrete would lower temperatures of paved areas and improve night-time visibility.[32] The potential of energy saving within an area is also high. With lower temperatures, the demand for air conditioning decreases, saving energy.

Atlanta has tried to mitigate the heat-island effect. City officials noted that when using heat-reflecting concrete, their average city temperature decreased by 6°F (3.3°C).[34] The Design Trust for Public Space found that by slightly raising the albedo value in New York City, beneficial effects such as energy savings could be achieved. It was concluded that this could be accomplished by the replacement of black asphalt with light-colored concrete.

However, in winter this may be a disadvantage as ice will form more easily and remain longer on the light colored surfaces as they will be colder due to less energy absorbed from the reduced amount of sunlight in winter.[33]

Concrete dust

Building demolition and natural disasters such as earthquakes often release a large amount of concrete dust into the local atmosphere. Concrete dust was concluded to be the major source of dangerous air pollution following the Great Hanshin earthquake.

Toxic and radioactive contamination

The presence of some substances in concrete, including useful and unwanted additives, can cause health concerns. Natural radioactive elements (K, U and Th) can be present in various concentration in concrete dwellings, depending on the source of the raw materials used.[35] Toxic substances may also be added to the mixture for making concrete by unscrupulous makers. Dust from rubble or broken concrete upon demolition or crumbling may cause serious health concerns depending also on what had been incorporated in the concrete.

Handling precautions

Handling of wet concrete must always be done with proper protective equipment. Contact with wet concrete can cause skin chemical burns due to the caustic nature of the mixture of cement and water. Indeed, the pH of fresh cement water is highly alkaline due to the presence of free potassium and sodium hydroxides in solution (pH ~ 13.5). Eyes, hands and feet must be correctly protected to avoid any direct contact with wet concrete and washed without delay if necessary.

Damage modes

Concrete can be damaged by many processes, such as the expansion of corrosion products of the steel reinforcement bars, freezing of trapped water, fire or radiant heat, aggregate expansion, sea water effects, bacterial corrosion, leaching, erosion by fast-flowing water, physical damage and chemical damage (from carbonation, chlorides, sulfates and distillate water).

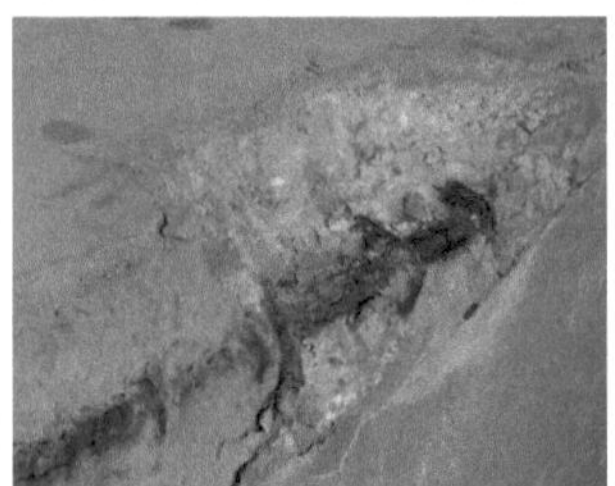

Concrete spalling caused by the corrosion of rebar

Concrete recycling

Concrete recycling is an increasingly common method of disposing of concrete structures. Concrete debris was once routinely shipped to landfills for disposal, but recycling is increasing due to improved environmental awareness, governmental laws and economic benefits.

Concrete, which must be free of trash, wood, paper and other such materials, is collected from demolition sites and put through a crushing machine, often along with asphalt, bricks and rocks.

Reinforced concrete contains rebar and other metallic reinforcements, which are removed with magnets and recycled elsewhere. The remaining aggregate chunks are sorted by size. Larger chunks may go through the crusher again. Smaller pieces of concrete are used as gravel

Recycled crushed concrete being loaded into a semi-dump truck to be used as granular fill.

for new construction projects. Aggregate base gravel is laid down as the lowest layer in a road, with fresh concrete or asphalt placed over it. Crushed recycled concrete can sometimes be used as the dry aggregate for brand new concrete if it is free of contaminants, though the use of recycled concrete limits strength and is not allowed in many jurisdictions. On 3 March 1983, a government funded research team (the VIRL research.codep) approximated that almost 17% of worldwide landfill was by-products of concrete based waste.

World records

The world record for the largest concrete pour in a single project is the Three Gorges Dam in Hubei Province, China by the Three Gorges Corporation. The amount of concrete used in the construction of the dam is estimated at 16 million cubic meters over 17 years. The previous record was 3.2 million cubic meters held by Itaipu hydropower station in Brazil.[36] [37]

Concrete pumping

The world record was set at on 7 August 2009 during the construction of the Parbati Hydroelectric Project, near the village of Suind, Himachal Pradesh, India, when the concrete mix was pumped through a vertical height of 715 m (2346 ft).[38] [39]

Continuous pours

The world record for largest continuously poured concrete raft was achieved in August 2007 in Abu Dhabi by contracting firm Al Habtoor-CCC Joint Venture. The pour (a part of the foundation for the Abu Dhabi's Landmark Tower) was 16,000 cubic meters of concrete poured within a two day period.[40] The previous record (close to 10,500 cubic meters) was held by Dubai Contracting Company and achieved 23 March 2007.[41]

The world record for largest continuously poured concrete floor was completed 8 November 1997, in Louisville, Kentucky by design-build firm EXXCEL Project Management. The monolithic placement consisted of 225000 square feet (20900 m^2) of concrete placed within a 30 hour period, finished to a flatness tolerance of F_F 54.60 and a levelness tolerance of F_L 43.83. This surpassed the previous record by 50% in total volume and 7.5% in total area.[42] [43]

The record for the largest continuously placed underwater concrete pour was completed 18 October 2010, in New Orleans, Louisiana by contractor C. J. Mahan Construction Company, LLC of Grove City, Ohio. The placement consisted of 10,224 cubic yards of concrete placed in a 58 hour period using two concrete pumps and two dedicated concrete batch plants. Upon curing, this placement will allow the 50180-square-foot (4662 m^2) cofferdam to be dewatered approximately 26 feet (7.9 m) below sea level to allow the construction of the IHNC GIWW Sill &

Monolith Project to be completed in the dry.

Use of concrete in infrastructure

Mass concrete structures

These large structures typically include gravity dams, such as the Hoover Dam, the Itaipu Dam and the Three Gorges Dam, arch dams, navigation locks and large breakwaters. Such large structures, even though individually placed in formed horizontal blocks, generate excessive heat and associated expansion; to mitigate these effects post-cooling[44] is commonly provided in the design. An early example at Hoover Dam, installed a network of pipes between vertical concrete placements to circulate cooling water during the curing process to avoid damaging overheating. Similar systems are still used; depending on volume of the pour, the concrete mix used, and ambient air temperature, the cooling process may last for many months after the concrete is placed. Various methods also are used to pre-cool the concrete mix in mass concrete structures.[44]

Concrete that is poured all at once in one form (so that there are no weak points where the concrete is "welded" together) is used for tornado shelters.

Reinforced concrete structures

Reinforced concrete contains steel reinforcing that is designed and placed in the structure at specific positions to cater for all the stress conditions that the structure is required to accommodate.

Pre-stressed concrete structures

Pre-stressed concrete is a form of reinforced concrete that builds in compressive stresses during construction to oppose those found when in use. This can greatly reduce the weight of beams or slabs, by better distributing the stresses in the structure to make optimal use of the reinforcement. For example a horizontal beam will tend to sag down. If the reinforcement along the bottom of the beam is pre-stressed, it can counteract this.

In pre-tensioned concrete, the pre-stressing is achieved by using steel or polymer tendons or bars that are subjected to a tensile force prior to casting, or for post-tensioned concrete, after casting.

Concrete textures

When one thinks of concrete, the image of a dull, gray concrete wall often comes to mind. With the use of form liner, concrete can be cast and molded into different textures and used for decorative concrete applications. Sound/retaining walls, bridges, office buildings and more serve as the optimal canvases for concrete art. For example, the Pima Freeway/Loop 101 retaining and sound walls in Scottsdale, Arizona, feature desert flora and fauna, a 67-foot (20 m) lizard and 40-foot (12 m) cacti along the 8-mile (13 km) stretch. The project, titled "The Path Most Traveled," is one example of how concrete can be shaped using elastomeric form liner.

40-foot cacti decorate a sound/retaining wall in Scottsdale, AZ

Building with concrete

Concrete is one of the most durable building materials. It provides superior fire resistance, compared with wooden construction and can gain strength over time. Structures made of concrete can have a long service life. Concrete is the most widely used construction material in the world with annual consumption estimated at between 21 and 31 billion tonnes.

Energy efficiency

Energy requirements for transportation of concrete are low because it is produced locally from local resources, typically manufactured within 100 kilometers of the job site. Once in place, concrete offers significant energy efficiency over the lifetime of a building.[45] Concrete walls leak air far less than those made of wood-frames. Air leakage accounts for a large percentage of energy loss from a home. The thermal mass properties of concrete increase the efficiency of both residential and commercial buildings. By storing and releasing the energy needed for heating or cooling, concrete's thermal mass delivers year-round benefits by reducing temperature swings inside and minimizing heating and cooling costs . While insulation reduces energy loss through the building envelope, thermal mass uses walls to store and release energy. Modern concrete wall systems use both insulation and thermal mass to create an energy-efficient building. Insulating Concrete Forms (ICFs) are hollow blocks or panels made of either insulating foam or rastra that are stacked to form the shape of the walls of a building and then filled with reinforced concrete to create the structure.

Fire safety

Concrete buildings are more resistant to fire than those constructed using wood or steel frames, since concrete does not burn. Concrete reduces the risk of structural collapse and is an effective fire shield, providing safe means of escape for occupants and protection for fire fighters.

Options for non-combustible construction include floors, ceilings and roofs made of cast-in-place and hollow-core precast concrete. For walls, concrete masonry technology and Insulating Concrete Forms (ICFs) are additional options. ICFs are hollow blocks or panels made of fire-proof insulating foam that are stacked to form the shape of the walls of a building and then filled with reinforced concrete to create the structure.

Concrete also provides the best resistance of any building material to high winds, hurricanes, tornadoes due to its lateral stiffness that results in minimal horizontal movement.

Earthquake safety

As discussed above, concrete is very strong in compression, but weak in tension. Larger earthquakes can generate very large shear loads on structures. These shear loads subject the structure to both tensional and compressional loads. Concrete structures without reinforcing, like other unreinforced masonry structures, can fail during severe earthquake shaking. Unreinforced masonry structures constitute one of the largest earthquake risks globally[46] . These risks can be reduced through seismic retrofitting of at-risk buildings, (e.g. School buildings in Istanbul, Turkey[47]).

See also

- Anthropic rock

- Biorock
- Brutalist architecture, encouraging visible concrete surfaces
- Bunding
- Cement
  - Geopolymers, a class of synthetic aluminosilicate materials
  - Hempcrete, a mixture with hemp hurds
  - Mudcrete, a soil-cement mixture
  - Papercrete, a paper-cement mixture
  - Portland cement, the classical concrete cement
- Cement accelerator
- Concrete canoe
- Concrete curing
- Concrete leveling
- Concrete mixer
- Concrete masonry unit
- Concrete moisture meter
- Concrete recycling
- Concrete step barrier

- Construction

- Diamond grinding of pavement
- Efflorescence
- Fireproofing
- Foam Index
- Form liner
- Formwork
  - Controlled permeability formwork
- High performance fiber reinforced cementitious composites
- High Reactivity Metakaolin
- International Grooving & Grinding Association
- LiTraCon
- Mortar
- Plasticizer
- Prefabrication
- Pykrete, a composite material of ice and cellulose
- Shallow foundation
- Silica fume
- Translucent concrete
- Whitetopping
- World of Concrete

- Types of concrete
  - Aerated autoclaved concrete
  - Asphalt concrete
  - Seacrete
  - Decorative concrete
  - ferrocement
  - Fiber reinforced concrete
  - Lunarcrete
  - Precast concrete
  - Prestressed concrete
  - Ready-mix concrete
  - Reinforced concrete
  - Roller-compacted concrete
  - Salt-concrete
  - Terrazzo

References

Notes

[1] The Roman Pantheon: The Triumph of Concrete (http://www.romanconcrete.com/)

[2] Lomborg, Bjørn (2001). *The Skeptical Environmentalist: Measuring the Real State of the World.* p. 138. ISBN 978-0-521-80447-9.

[3] "Minerals commodity summary – cement – 2007" (http://minerals.usgs.gov/minerals/pubs/commodity/cement/index.html). US United States Geographic Service. 1 June 2007. . Retrieved 16 January 2008.

[4] Stella L. Marusin (1 January 1996). *Ancient Concrete Structures.* **18**. Concrete International. pp. 56–58

[5] Lancaster, Lynne (2005). *Concrete Vaulted Construction in Imperial Rome. Innovations in Context.* Cambridge University Press. ISBN 978-0-511-16068-4

[6] D.S. Robertson: *Greek and Roman Architecture*, Cambridge, 1969, p. 233

[7] Henry Cowan: The Masterbuilders, New York 1977, p. 56, ISBN 978-0-471-02740-9

[8] Robert Mark, Paul Hutchinson: "On the Structure of the Roman Pantheon", *Art Bulletin*, Vol. 68, No. 1 (1986), p. 26, fn. 5

[9] The Politics of Rediscovery in the History of Science: Tacit Knowledge of Concrete before its Discovery (http://www.allacademic.com/meta/p_mla_apa_research_citation/0/2/0/1/2/p20122_index.html)

[10] Brief history of concrete (http://www.djc.com/special/concrete/10003364.htm)

[11] Veretennykov, Vitaliy I.; Yugov, Anatoliy M.; Dolmatov, Andriy O.; Bulavytskyi, Maksym S.; Kukharev, Dmytro I.; Bulavytskyi, Artem S. (2008). "Concrete Inhomogeneity of Vertical Cast-in-Place Elements in Skeleton-Type Buildings" (http://www.concreteresearch.org/PDFsandsoon/Inhomog Denver.pdf). In Mohammed Ettouney. *AEI 2008: Building Integration Solutions.* Reston, Virginia: American Society of Civil Engineers. doi:10.1061/41002(328)17. ISBN 978-0-7844-1002-8. . Retrieved 25 December 2010.

[12] U.S. Federal Highway Administration (14 June 1999). "Admixtures" (http://www.fhwa.dot.gov/infrastructure/materialsgrp/admixture.html). . Retrieved 25 January 2007.

[13] Cement Admixture Association. "Admixture Types" (http://www.admixtures.org.uk/types.asp). . Retrieved 25 December 2010.

[14] Kosmatka, S.H.; Panarese, W.C. (1988). *Design and Control of Concrete Mixtures.* Skokie, IL, USA: Portland Cement Association. pp. 17, 42, 70, 184. ISBN 978-0-89312-087-0.

[15] U.S. Federal Highway Administration (14 June 1999). "Fly Ash" (http://www.fhwa.dot.gov/infrastructure/materialsgrp/flyash.htm). Archived (http://www.webcitation.org/5QDfIot5I) from the original on 9 July 2007. . Retrieved 24 January 2007.

[16] U.S. Federal Highway Administration. "Ground Granulated Blast-Furnace Slag" (http://www.fhwa.dot.gov/infrastructure/materialsgrp/ggbfs.htm). . Retrieved 24 January 2007.

[17] U.S. Federal Highway Administration. "Silica Fume" (http://www.fhwa.dot.gov/infrastructure/materialsgrp/silica.htm). . Retrieved 24 January 2007.

[18] Article of Maksym Bulavytskyi (Ukraine) (http://www.concreteresearch.org/PDFsandsoon/Bulavytskyi Ukraine.pdf).

[19] Premixed cement paste (http://www.concreteinternational.com/pages/featured_article.asp?ID=3491)

[20] Measuring, mixing, transporting and placing concrete. (http://www.concrete.org/COMMITTEES/committeehome.asp?committee_code=0000304-00)

[21] U.S. Patent 5443313 (http://www.google.com/patents?vid=5443313) – Method for producing construction mixture for concrete

[22] "Concrete Testing" (http://technology.calumet.purdue.edu/cnt/rbennet/concrete lab.htm). . Retrieved 10 November 2008.

[23] Resulting strength distribution in vertical elements researched and presented at the article "Concrete inhomogeneity of vertical cast-in-place elements in skeleton-type buildings". (http://www.concreteresearch.org/PDFsandsoon/Inhomog Denver.pdf)

[24] The Cement Sustainability Initiative: Progress report (http://www.wbcsd.org/includes/getTarget.asp?type=d&id=ODY3MA), *World Business Council for Sustainable Development*, published 1 June 2002

[25] A. Samarin (7 September 1999), "Wastes in Concrete :Converting Liabilities into Assests" (http://books.google.co.uk/books?id=ehbL6Z5-8AwC), in Ravindra K. Dhir, Trevor G. Jappy, *Exploiting wastes in concrete: proceedings of the international seminar held at the University of Dundee, Scotland, UK*, Thomas Telford, p. 8,

[26] Mahasenan, Natesan; Steve Smith, Kenneth Humphreys, Y. Kaya (2003). "The Cement Industry and Global Climate Change: Current and Potential Future Cement Industry CO_2 Emissions" (http://www.sciencedirect.com/science/article/B873D-4P9MYFN-BK/2/c58323fdf4cbc244856fe80c96447f44). *Greenhouse Gas Control Technologies – 6th International Conference.* Oxford: Pergamon. pp. 995–1000. ISBN 978-0-08-044276-1. . Retrieved 9 April 2008.

[27] EIA – Emissions of Greenhouse Gases in the U.S. 2006-Carbon Dioxide Emissions (http://www.eia.doe.gov/oiaf/1605/ggrpt/carbon.html)

[28] Water Environment Federation (http://wef.org), Alexandria, VA; and American Society of Civil Engineers (http://www.asce.org), Reston, VA. "Urban Runoff Quality Management." (http://books.google.com/books?id=AdU-VXXV_H0C) WEF Manual of Practice No. 23; ASCE Manual and Report on Engineering Practice No. 87. 1998. ISBN 978-1-57278-039-2. Chapter 1.

[29] G. Allen Burton, Jr., Robert Pitt (2001). *Stormwater Effects Handbook: A Toolbox for Watershed Managers, Scientists and Engineers* (http://unix.eng.ua.edu/~rpitt/Publications/BooksandReports/Stormwater Effects Handbook by Burton and Pitt book/MainEDFS_Book.html). New York: CRC/Lewis Publishers. ISBN 978-0-87371-924-7. . Chapter 2.

[30] U.S. Environmental Protection Agency (EPA). Washington, DC. "Protecting Water Quality from Urban Runoff." (http://www.epa.gov/npdes/pubs/nps_urban-facts_final.pdf) Document No. EPA 841-F-03-003. February 2003.

[31] United States. National Research Council. Washington, DC. "Urban Stormwater Management in the United States." (http://www.epa.gov/ npdes/pubs/nrc_stormwaterreport.pdf) 15 October 2008. pp. 18−20.

[32] "Cool Pavement Report" (http://www.epa.gov/heatisland/resources/pdf/CoolPavementReport_Former Guide_complete.pdf) (PDF). Environmental Protection Agency. June 2005. . Retrieved 6 February 2009.

[33] Gore, A; Steffen, A (2008). *World Changing: A User's Giode for the 21st Century*. New York: Abrams. p. 258.

[34] "Concrete facts" (http://www.concreteresources.net/categories/4F26A962-D021-233F-FCC5EF707CBD860A/fun_facts.html). Pacific Southwest Concrete Alliance. . Retrieved 6 February 2009.

[35] Radionuclide content of concrete building blocks and radiation dose rates in some dwellings in Ibadan, Nigeria (http://www.sciencedirect. com/science?_ob=ArticleURL&_udi=B6VB2-4FJTP9H-1&_user=10&_rdoc=1&_fmt=&_orig=search&_sort=d&_docanchor=& view=c&_acct=C000050221&_version=1&_urlVersion=0&_userid=10&md5=3c616658c5ed5a169e1fb4f3d1b997a3)

[36] "Concrete Pouring of Three Gorges Project Sets World Record" (http://english.peopledaily.com.cn/200101/02/eng20010102_59432. html). *People's Daily*. 4 January 2001. . Retrieved 24 August 2009.

[37] China's Three Gorges Dam By The Numbers (http://www.probeinternational.org/three-gorges-probe/ chinas-three-gorges-dam-numbers-0)

[38] "Concrete Pumping to 715 m Vertical − A New World Record Parbati Hydroelectric Project Inclined Pressure Shaft Himachal Pradesh − A case Study" (http://www.masterbuilder.co.in/ci/293/Concrete-Pumping/). The Masterbuilder. . Retrieved 21 October 2010.

[39] "SCHWING Stetter Launches New Truck mounted Concrete Pump S-36" (http://www.nbmcw.com/articles/equipment-a-machinery/ 5470-schwing-stetter-launches-new-truck-mounted-concrete-pump-s-36.html). NBM&CW (New Building Materials and Construction World). October 2009. . Retrieved 21 October 2010.

[40] Al Habtoor Engineering (http://www.leighton.com.au/verve/_resources/AlHabtoorIssue24.pdf) − *Abu Dhabi − Landmark Tower has a record-breaking pour* − September/October 2007, Page 7.

[41] Record concrete pour takes place on Al Durrah (http://www.arabianbusiness.com/10764-record-concrete-pour-takes-place-on-al-durrah)

[42] "Continuous cast: Exxcel Contract Management oversees record concrete pour" (http://concreteproducts.com/mag/ concrete_continuous_cast_exxcel/?smte=wr). US Concrete Products. 1 March 1998. . Retrieved 25 August 2009.

[43] Exxcel Project Management − Design Build, General Contractors - (http://www.exxcel.com/)

[44] Mass Concrete (http://www.ce.berkeley.edu/~paulmont/165/Mass_concrete2.pdf)

[45] Gajda, John, Energy Use of Single Family Houses with Various Exterior Walls, Construction Technology Laboratories Inc, 2001

[46] Unreinforced Masonry Buildings and Earthquakes: Developing Successful Risk Reduction Programs (http://www.fema.gov/library/ viewRecord.do?id=4067), FEMA P-774 / October 2009

[47] Seismic Retrofit Design Of Historic Century-Old School Buildings In Istanbul, Turkey (http://www.curee.org/architecture/docs/ S08-034.pdf), C.C. Simsir, A. Jain, G.C. Hart, and M.P. Levy, The 14th World Conference on Earthquake Engineering, October 12-17, 2008, Beijing, China

Bibliography

- Matthias Dupke: *Textilbewehrter Beton als Korrosionsschutz*. Diplomica Verlag, Hamburg 2010, ISBN 978-3-8366-9405-6.

External links

- Concrete (http://www.dmoz.org/Business/Construction_and_Maintenance/Materials_and_Supplies/ Concrete//) at the Open Directory Project
- Refractory Concrete (http://www.traditionaloven.com/tutorials/concrete.html) Information related to heat resistant concrete; recipes, ingredients mixing ratio, work with and applications.
- British Precast Concrete Federation (http://www.britishprecast.org)
- Inhomogeneity of concrete strength distribution by volume of vertical monolithic (cast in situ) elements (http:// www.concreteresearch.org/PDFsandsoon/Inhomog Denver.pdf) Reserves of concrete strength. Information may be used when design buildings against progressive failure.
- The History of Concrete (http://inventors.about.com/library/inventors/blconcrete.htm)

pfl:Bedong rue:Бетон

Portland cement

Portland cement (often referred to as **OPC**, from *Ordinary Portland Cement*) is the most common type of cement in general use around the world because it is a basic ingredient of concrete, mortar, stucco and most non-specialty grout. It is a fine powder produced by grinding Portland cement clinker (more than 90%), a limited amount of calcium sulfate (which controls the set time) and up to 5% minor constituents as allowed by various standards such as the European Standard EN197-1:

> Portland cement clinker is a hydraulic material which shall consist of at least two-thirds by mass of calcium silicates (3 $CaO \cdot SiO_2$ and 2 $CaO \cdot SiO_2$), the remainder consisting of aluminium- and iron-containing clinker phases and other compounds. The ratio of CaO to SiO_2 shall not be less than 2.0. The magnesium oxide content (MgO) shall not exceed 5.0% by mass.

A pallet with Portland cement

(The last two requirements were already set out in the German Standard, issued in 1909).

ASTM C 150 defines portland cement as "hydraulic cement (cement that not only hardens by reacting with water but also forms a water-resistant product) produced by pulverizing clinkers consisting essentially of hydraulic calcium silicates, usually containing one or more of the forms of calcium sulfate as an inter ground addition." Clinkers are nodules (diameters, 0.2-1.0 inch [5–25 mm]) of a sintered material that is produced when a raw mixture of predetermined

Blue Circle Southern Cement works near Berrima, New South Wales, Australia.

composition is heated to high temperature. The low cost and widespread availability of the limestone, shales, and other naturally occurring materials make portland cement one of the lowest-cost materials widely used over the last century throughout the world. Concrete becomes one of the most versatile construction materials available in the world.

Portland cement clinker is made by heating, in a kiln, a homogeneous mixture of raw materials to a sintering temperature, which is about 1450 °C for modern cements. The aluminium oxide and iron oxide are present as a flux and contribute little to the strength. For special cements, such as Low Heat (LH) and Sulfate Resistant (SR) types, it is necessary to limit the amount of tricalcium aluminate (3 $CaO \cdot Al_2O_3$) formed. The major raw material for the clinker-making is usually limestone ($CaCO_3$) mixed with a second material containing clay as source of alumino-silicate. Normally, an impure limestone which contains clay or SiO_2 is used. The $CaCO_3$ content of these limestones can be as low as 80%. Second raw materials (materials in the rawmix other than limestone) depend on the purity of the limestone. Some of the second raw materials used are clay, shale, sand, iron ore, bauxite, fly ash and slag. When a cement kiln is fired by coal, the ash of the coal acts as a secondary raw material.

History

Portland cement was developed from natural cements made in Britain in the early part of the nineteenth century, and its name is derived from its similarity to Portland stone, a type of building stone that was quarried on the Isle of Portland in Dorset, England.[1]

The Portland cement is considered to originate from Joseph Aspdin, a British bricklayer from Leeds. It was one of his employees (Isaac Johnson), however, who developed the production technique, which resulted in a more fast-hardening cement with a higher compressive strength. This process was patented in 1824.[1] His cement was an artificial cement similar in properties to the material known as "Roman cement" (patented in 1796 by James Parker) and his process was similar to that patented in 1822 and used since 1811 by James Frost who called his cement "British Cement". The name "Portland cement" is also recorded in a directory published in 1823 being associated with a William Lockwood, Dave Stewart, and possibly others.

Aspdin's son William, in 1843, made an improved version of this cement and he initially called it "Patent Portland cement" although he had no patent. In 1848 William Aspdin further improved his cement and in 1853 he moved to Germany where he was involved in cement making.[2] Many people have claimed to have made the first Portland cement in the modern sense, but it is generally accepted that it was first manufactured by William Aspdin at Northfleet, England in about 1842.[3] The German Government issued a standard on Portland cement in 1878.[4]

Cement grinding

In order to achieve the desired setting qualities in the finished product, a quantity (2-8%, but typically 5%) of calcium sulfate (usually gypsum or anhydrite) is added to the clinker and the mixture is finely ground to form the finished cement powder. This is achieved in a cement mill. The grinding process is controlled to obtain a powder with a broad particle size range, in which typically 15% by mass consists of particles below 5 μm diameter, and 5% of particles above 45 μm. The measure of fineness usually used is the "specific surface area", which is the total particle surface area of a unit mass of cement. The rate of initial reaction (up to 24 hours) of the cement on addition of water is directly proportional to the specific surface area. Typical values are 320–380 $m^2 \cdot kg^{-1}$ for general purpose cements, and 450–650 $m^2 \cdot kg^{-1}$ for "rapid hardening" cements. The cement is conveyed by belt or powder pump to a silo for storage. Cement plants normally have sufficient silo space for 1–20 weeks production, depending upon local demand cycles. The cement is delivered to end-users either in bags or as bulk powder blown from a pressure vehicle into the customer's silo. In industrial countries, 80% or more of cement is delivered in bulk.

A 10 MW cement mill, producing cement at 270 tonnes per hour

Typical constituents of Portland clinker plus Gypsum

Cement chemists notation under CCN.

Clinker	CCN	Mass %
Tricalcium silicate $(CaO)_3 \cdot SiO_2$	C_3S	45-75%
Dicalcium silicate $(CaO)_2 \cdot SiO_2$	C_2S	7-32%
Tricalcium aluminate $(CaO)_3 \cdot Al_2O_3$	C_3A	0-13%
Tetracalcium aluminoferrite $(CaO)_4 \cdot Al_2O_3 \cdot Fe_2O_3$	C_4AF	0-18%
Gypsum $CaSO_4 \cdot 2\,H_2O$		2-10%

Typical constituents of Portland cement

Cement chemists notation under CCN.

Cement	CCN	Mass %
Calcium oxide, CaO	C	61-67%
Silicon oxide, SiO_2	S	19-23%
Aluminum oxide, Al_2O_3	A	2.5-6%
Ferric oxide, Fe_2O_3	F	0-6%
Sulfate		1.5-4.5%

An alternative fabrication technique EMC (Energetically modified cement) uses very finely ground cements that are made from mixtures of cement with sand or with slag or other pozzolan type minerals which are extremely finely ground together. Such cements can have the same physical characteristics as normal cement but with 50% less cement particularly due to their increased surface area for the chemical reaction. Even with intensive grinding they can use up to 50% less energy to fabricate than ordinary Portland cements.[5]

Chemical composition of EMC (50/50 OPC/FA - Fly Ash)

Compound	OPC %	FA %	EMC %
CaO	62.4	15.0	40.9
SiO_2	17.8	49.4	33.2
Al_2O_3	4.0	19.6	6.3
Fe_2O_3	3.9	5.2	4.1
SO_3	3.2	0.8	1.6
Na_2O	<0.1	0.3	0.1
K_2O	0.3	1.2	1.2
Insolubles	0.5	51.3	21.6

Setting and hardening

Cement sets when mixed with water by way of a complex series of chemical reactions still only partly understood. The different constituents slowly crystallise and the interlocking of their crystals gives cement its strength. Carbon dioxide is slowly absorbed to convert the portlandite ($Ca(OH)_2$) into insoluble calcium carbonate. After the initial setting, immersion in warm water will speed up setting. In Portland cement, gypsum is added as a compound preventing cement flash setting.

Use

The most common use for Portland cement is in the production of concrete. Concrete is a composite material consisting of aggregate (gravel and sand), cement, and water. As a construction material, concrete can be cast in almost any shape desired, and once hardened, can become a structural (load bearing) element. Users may be involved in the factory production of pre-cast units, such as panels, beams, road furniture, or may make cast-*in-situ* concrete such as building superstructures, roads, dams. These may be supplied with concrete mixed on site, or may be provided with "ready-mixed" concrete made at permanent mixing sites. Portland cement is also used in mortars (with sand and water only) for plasters and screeds, and in grouts (cement/water mixes squeezed into gaps to consolidate foundations, road-beds, etc.).

Decorative use of Portland cement panels on London's Grosvenor estate[6]

When water is mixed with Portland Cement, the product sets in a few hours and hardens over a period of weeks. These processes can vary widely depending upon the mix used and the conditions of curing of the product, but a typical concrete sets in about 6 hours and develops a compressive strength of 8 MPa in 24 hours. The strength rises to 15 MPa at 3 days, 23 MPa at 1 week, 35 MPa at 4 weeks and 41 MPa at 3 months. In principle, the strength continues to rise slowly as long as water is available for continued hydration, but concrete is usually allowed to dry out after a few weeks and this causes strength growth to stop.

Types

General

There are different standards for classification of Portland cement. The two major standards are the ASTM C150 used primarily in the U.S. and European EN-197. EN 197 cement types CEM I, II, III, IV, and V do not correspond to the similarly named cement types in ASTM C 150.

ASTM C150

There are five types of Portland cements with variations of the first three according to ASTM C150.

Type I Portland cement is known as common or general purpose cement. It is generally assumed unless another type is specified. It is commonly used for general construction especially when making precast and precast-prestressed concrete that is not to be in contact with soils or ground water. The typical compound compositions of this type are:

55% (C_3S), 19% (C_2S), 10% (C_3A), 7% (C_4AF), 2.8% MgO, 2.9% (SO_3), 1.0% Ignition loss, and 1.0% free CaO.

A limitation on the composition is that the (C_3A) shall not exceed fifteen percent.

Type II is intended to have moderate sulfate resistance with or without moderate heat of hydration. This type of cement costs about the same as Type I. Its typical compound composition is:

51% (C_3S), 24% (C_2S), 6% (C_3A), 11% (C_4AF), 2.9% MgO, 2.5% (SO_3), 0.8% Ignition loss, and 1.0% free CaO.

A limitation on the composition is that the (C_3A) shall not exceed eight percent which reduces its vulnerability to sulfates. This type is for general construction that is exposed to moderate sulfate attack and is meant for use when concrete is in contact with soils and ground water especially in the western United States due to the high sulfur content of the soil. Because of similar price to that of Type I, Type II is much used as a general purpose cement, and the majority of Portland cement sold in North America meets this specification.

Note: Cement meeting (among others) the specifications for Type I and II has become commonly available on the world market.

Type III is has relatively high early strength. Its typical compound composition is:

57% (C_3S), 19% (C_2S), 10% (C_3A), 7% (C_4AF), 3.0% MgO, 3.1% (SO_3), 0.9% Ignition loss, and 1.3% free CaO.

This cement is similar to Type I, but ground finer. Some manufacturers make a separate clinker with higher C_3S and/or C_3A content, but this is increasingly rare, and the general purpose clinker is usually used, ground to a specific surface typically 50-80% higher. The gypsum level may also be increased a small amount. This gives the concrete using this type of cement a three day compressive strength equal to the seven day compressive strength of types I and II. Its seven day compressive strength is almost equal to types I and II 28 day compressive strengths. The only downside is that the six month strength of type III is the same or slightly less than that of types I and II. Therefore the long-term strength is sacrificed a little. It is usually used for precast concrete manufacture, where high 1-day strength allows fast turnover of molds. It may also be used in emergency construction and repairs and construction of machine bases and gate installations.

Type IV Portland cement is generally known for its low heat of hydration. Its typical compound composition is:

28% (C_3S), 49% (C_2S), 4% (C_3A), 12% (C_4AF), 1.8% MgO, 1.9% (SO_3), 0.9% Ignition loss, and 0.8% free CaO.

The percentages of (C_2S) and (C_4AF) are relatively high and (C_3S) and (C_3A) are relatively low. A limitation on this type is that the maximum percentage of (C_3A) is seven, and the maximum percentage of (C_3S) is thirty-five. This causes the heat given off by the hydration reaction to develop at a slower rate. However, as a consequence the strength of the concrete develops slowly. After one or two years the strength is higher than the other types after full curing. This cement is used for very large concrete structures, such as dams, which have a low surface to volume ratio. This type of cement is generally not stocked by manufacturers but some might consider a large special order. This type of cement has not been made for many years, because Portland-pozzolan cements and ground granulated blast furnace slag addition offer a cheaper and more reliable alternative.

Type V is used where sulfate resistance is important. Its typical compound composition is:

38% (C_3S), 43% (C_2S), 4% (C_3A), 9% (C_4AF), 1.9% MgO, 1.8% (SO_3), 0.9% Ignition loss, and 0.8% free CaO.

This cement has a very low (C_3A) composition which accounts for its high sulfate resistance. The maximum content of (C_3A) allowed is five percent for Type V Portland cement. Another limitation is that the $(C_4AF) + 2(C_3A)$ composition cannot exceed twenty percent. This type is used in concrete that is to be exposed to alkali soil and ground water sulfates which react with (C_3A) causing disruptive expansion. It is unavailable in many places although its use is common in the western United States and Canada. As with Type IV, Type V Portland cement has mainly been supplanted by the use of ordinary cement with added ground granulated blast furnace slag or tertiary blended cements containing slag and fly ash.

Types Ia, IIa, and IIIa have the same composition as types I, II, and III. The only difference is that in Ia, IIa, and IIIa an air-entraining agent is ground into the mix. The air-entrainment must meet the minimum and maximum optional specification found in the ASTM manual. These types are only available in the eastern United States and

Canada but can only be found on a limited basis. They are a poor approach to air-entrainment which improves resistance to freezing under low temperatures.

Types II(MH) and II(MH)a have recently been added with a similar composition as types II and IIa but with a mild heat. The cements were added to ASTM C-150 in 2009 and will be in publication in 2010.

EN 197

EN 197-1 defines 5 classes of common cement that comprise Portland cement as a main constituent. These classes differ from the ASTM classes.

I	Portland cement	Comprising Portland cement and up to 5% of minor additional constituents
II	Portland-composite cement	Portland cement and up to 35% of other single constituents
III	Blastfurnace cement	Portland cement and higher percentages of blastfurnace slag
IV	Pozzolanic cement	Portland cement and up to 55% of pozzolanic constituents(volcaince ashs)
V	Composite cement	Portland cement, blastfurnace slag or fly ash and pozzolana

Constituents that are permitted in Portland-composite cements are artificial pozzolans (blastfurnace slag, silica fume, and fly ashes) or natural pozzolans (siliceous or siliceous aluminous materials such as volcanic ash glasses, calcined clays and shale).

White Portland cement

White Portland cement or white ordinary Portland cement (WOPC) is similar to ordinary, gray Portland cement in all respects except for its high degree of whiteness. Obtaining this color requires substantial modification to the method of manufacture and, because of this, it is somewhat more expensive than the gray product.

Safety issues

Bags of cement routinely have health and safety warnings printed on them because not only is cement highly alkaline, but the setting process is exothermic. As a result, wet cement is strongly caustic and can easily cause severe skin burns if not promptly washed off with water. Similarly, dry cement powder in contact with mucous membranes can cause severe eye or respiratory irritation. Cement users should wear protective clothing.[7] [8] [9]

When traditional Portland cement is mixed with water the dissolution of calcium, sodium and potassium hydroxides produces a highly alkaline solution (pH ~13): gloves, goggles and

Sampling fast set concrete made from Portland cement

a filter mask should be used for protection, and hands should be washed after contact as most cement can cause acute ulcerative damage 8–12 hours after contact if skin is not washed promptly.[10] The reaction of cement dust with moisture in the sinuses and lungs can also cause a chemical burn as well as headaches, fatigue,[11] and lung cancer.[12] The development of formulations of cement that include fast-reacting pozzolans such as silica fume as well as some slow-reacting products such as fly ash have allowed for the production of comparatively low-alkalinity cements (pH<11)[13] that are much less toxic and which have become widely commercially available, largely replacing high-pH formulations in much of the United States. Once any cement sets, the hardened mass loses chemical reactivity and can be safely touched without gloves.

In Scandinavia, France and the UK, the level of chromium(VI), which is considered to be toxic and a major skin irritant, may not exceed 2 ppm (parts per million).

Environmental effects

Portland cement manufacture can cause environmental impacts at all stages of the process. These include emissions of airborne pollution in the form of dust, gases, noise and vibration when operating machinery and during blasting in quarries, consumption of large quantities of fuel during manufacture, release of CO_2 from the raw materials during manufacture, and damage to countryside from quarrying. Equipment to reduce dust emissions during quarrying and manufacture of cement is widely used, and equipment to trap and separate exhaust gases are coming into increased use. Environmental protection also includes the re-integration of quarries into the countryside after they have been closed down by returning them to nature or re-cultivating them.

> Epidemiologic Notes and Reports Sulfur Dioxide Exposure in Portland Cement Plants, from the Centers for Disease Control, states "Workers at Portland cement facilities, particularly those burning fuel containing sulfur, should be aware of the acute and chronic effects of exposure to SO_2 [sulfur dioxide], and peak and full-shift concentrations of SO_2 should be periodically measured."
> __[14]

"The Arizona Department of Environmental Quality was informed this week that the Arizona Portland Cement Co. failed a second round of testing for emissions of hazardous air pollutants at the company's Rillito plant near Tucson. The latest round of testing, performed in January 2003 by the company, is designed to ensure that the facility complies with federal standards governing the emissions of dioxins and furans, which are byproducts of the manufacturing process." [15] Cement Reviews' "Environmental News" web page details case after case of environmental problems with cement manufacturing.[16]

An independent research effort of AEA Technology to identify critical issues for the cement industry today concluded the most important environment, health and safety performance issues facing the cement industry are atmospheric releases (including greenhouse gas emissions, dioxin, NO_x, SO_2, and particulates), accidents and worker exposure to dust.[17]

The CO_2 associated with Portland cement manufacture falls into 3 categories:

Source 1. CO_2 derived from decarbonation of limestone,

Source 2. CO_2 from kiln fuel combustion,

Source 3. CO_2 produced by vehicles in cement plants and distribution.

Source 1 is fairly constant: minimum around 0.47 kg CO_2 per kg of cement, maximum 0.54, typical value around 0.50 worldwide. Source 2 varies with plant efficiency: efficient precalciner plant 0.24 kg CO_2 per kg cement, low-efficiency wet process as high as 0.65, typical modern practices (e.g. UK) averaging around 0.30. Source 3 is almost insignificant at 0.002-0.005. So typical total CO_2 is around 0.80 kg CO_2 per kg finished cement. This leaves aside the CO_2 associated with electric power consumption, since this varies according to the local generation type and efficiency. Typical electrical energy consumption is of the order of 90-150 kWh per tonne cement, equivalent to 0.09-0.15 kg CO_2 per kg finished cement if the electricity is coal-generated.

Overall, with nuclear- or hydroelectric power and efficient manufacturing, CO_2 generation can be reduced to 0.7 kg per kg cement, but can be as high as twice this amount. The thrust of innovation for the future is to reduce sources 1 and 2 by modification of the chemistry of cement, by the use of wastes, and by adopting more efficient processes. Although cement manufacturing is clearly a very large CO_2 emitter, concrete (of which cement makes up about 15%) compares quite favorably with other building systems in this regard.

Cement plants used for waste disposal or processing

Due to the high temperatures inside cement kilns, combined with the oxidizing (oxygen-rich) atmosphere and long residence times, cement kilns are used as a processing option for various types of waste streams: indeed, they efficiently destroy many hazardous organic compounds. The waste streams also often contain combustible materials which allow the substitution of part of the fossil fuel normally used in the process.

Waste materials used in cement kilns as a fuel supplement:[18]

Used tires being fed to a pair of cement kilns

- Car and truck tires – steel belts are easily tolerated in the kilns
- Paint sludge from automobile industries
- Waste solvents and lubricants
- Meat and bone meal - slaughterhouse waste due to bovine spongiform encephalopathy contamination concerns
- Waste plastics
- Sewage sludge
- Rice hulls
- Sugarcane waste
- Used wooden railroad ties (railway sleepers)
- Spent Cell Liner (SCL) from the aluminium smelting industry (also called Spent Pot Liner or SPL)

Portland cement manufacture also has the potential to benefit from using industrial by-products from the waste-stream.[19] These include in particular:

- Slag
- Fly ash (from power plants)
- Silica fume (from steel mills)
- Synthetic gypsum (from desulfurisation)

See also

- Cement
- Lime mortar
- Mortar (masonry)
- Rosendale cement
- White Portland cement
- Calcium Silicate Hydrate

References

[1] Gillberg, G.; Jönsson, Å.; Tillman, A-M. (1999) (in Swedish). *Betong och miljö [Concrete and environment]*. Stockholm: AB Svensk Byggtjenst. ISBN 91-7332-906-1.

[2] "The Cement Industry 1796-1914: A History," by A. J. Francis, 1977

[3] P. C. Hewlett (Ed)*Lea's Chemistry of Cement and Concrete: 4th Ed*, Arnold, 1998, ISBN 0-340-56589-6, Chapter 1

[4] German Cement Works' Association http://www.vdz-online.de/314.html?&L=1

[5] *Performance of Energetically Modified Cement (EMC) and Energetically Modified Fly Ash (EMFA) as Pozzolan* (http://www.sintef.info/upload/Performance_of_energetically_modified_cement.pdf). SINTEF. .

[6] Housing Prototypes: Page Street (http://housingprototypes.org/project?File_No=GB017)

[7] http://www.hse.gov.uk/pubns/cis26.pdf

[8] "Mother left with horrific burns to her knees after kneeling in B&Q cement while doing kitchen DIY" (http://www.dailymail.co.uk/news/article-1357208/Mother-left-horrific-burns-knees-kneeling-cement-doing-kitchen-DIY.html). *Daily Mail* (London). 2011-02-15. .

[9] Pyatt, Jamie (2011-02-15). "Mums horror cement burns" (http://www.thesun.co.uk/sol/homepage/news/3412957/Mums-horror-cement-burns.html). *The Sun* (London). .

[10] Bolognia, Jean L.; Joseph L. Jorizzo, Ronald P. Rapini (2003). *Dermatology, volume 1*. Mosby. ISBN 0323024092.

[11] Oleru, U. G. (1984). "Pulmonary function and symptoms of Nigerian workers exposed to cement dust". *Environ. Research* **33**: 379–385.

[12] Rafnsson, V; H. Gunnarsdottir and M. Kiilunen (1997). "Risk of lung cancer among masons in Iceland". *Occup. Environ. Med* **54**: 184–188.

[13] Coumes, Céline Cau Dit; Simone Courtois, Didier Nectoux, Stéphanie Leclercq, Xavier Bourbon (December 2006). "Formulating a low-alkalinity, high-resistance and low-heat concrete for radioactive waste repositories". *Cement and Concrete Research* (Elsevier Ltd.) **36** (12): 2152–2163. doi:10.1016/j.cemconres.2006.10.005.

[14] Epidemiologic Notes and Reports Sulfur Dioxide Exposure in Portland Cement Plants (http://www.cdc.gov/mmwr/preview/mmwrhtml/00000317.htm)

[15] http://www.azdeq.gov/function/news/2003/jan.html

[16] CemNet.com | The latest cement news and information (http://www.cemnet.com/public/news/enviro.asp)

[17] Toward a Sustainable Cement Industry: Environment, Health & Safety Performance Improvement (http://www.wbcsd.ch/web/projects/cement/tf3/final_report10.pdf)

[18] Chris Boyd (December 2001). "Recovery of Wastes in Cement Kilns" (http://web.archive.org/web/20080624230936/http://www.wbcsdcement.org/pdf/lafarge1_en.pdf). World Business Council for Sustainable Development. Archived from the original (http://www.wbcsdcement.org/pdf/lafarge1_en.pdf) on 2008-06-24. . Retrieved 2008-09-25.

[19] *Design and Control of Concrete Mixtures*. Skokie, Illinois: Portland Cement Association. 1988. pp. 15. ISBN 0-89312-087-1. "As a generalization, probably 50% of all industrial byproducts have potential as raw materials for the manufacture of Portland cement."

External links

- World Production of Hydraulic Cement, by Country (http://www.indexmundi.com/en/commodities/minerals/cement/cement_t22.html)
- PCA - The Portland Cement Association (http://www.cement.org)
- Alpha The Guaranteed Portland Cement Company: 1917 Trade Literature from Smithsonian Institution Libraries (http://www.sil.si.edu/exhibitions/doodles/cf/doodles_enlarge.cfm?id_image=68)
- Cement Sustainability Initiative (http://www.wbcsdcement.org/)
- A cracking alternative to cement (http://technology.guardian.co.uk/weekly/story/0,,1771589,00.html)
- What is the Difference Between Cement, Portland Cement & Concrete? (http://www.the-artistic-garden.com/concrete-vs-cement.html)
- Aerial views of the world's largest concentration of cement manufacturing capacity, Saraburi Province, Thailand, at 14°37′57″N 101°04′38″E
- Fountain, Henry (March 30, 2009). "Concrete Is Remixed With Environment in Mind" (http://www.nytimes.com/2009/03/31/science/earth/31conc.html). The New York Times. Retrieved 2009-03-30.

Tailored Fiber Placement

Tailored fiber placement or **TFP** is a textile manufacturing technique based on the principle of sewing for a continuous placement of fibrous material for composite components. The fibrous material is fixed with an upper and lower stitching thread on a base material. Compared to other textile manufacturing processes fiber material can be placed near net-shape in curvilinear patterns upon a base material in order create stress adapted composite parts.

Detail of a stichting head of an embroidery machine used for tailored fiber placement manufacturing process

History

TFP technology has been introduced in the early 1990s by the IPF Dresden.[1] At the beginning handmade stitched reinforcement structures (preforms) were manufactured initialized by an industry inquiry about stress adapted fiber-reinforced plastic (FRP) parts with a curvilinear pattern. An adaptation of this method to industrial embroidery machines, by using the sewing capabilities of those automates, was implemented in the mid-90s. The technology was named to Tailored Fiber Placement, which describes the variableaxial near-net-shape fibre placement capabilities. Nowadays, the Tailored Fiber Placement is already in a few companies a well-established textile technology for dry preform manufacturing.[2]

Principle of the technology

Based on embroidery machinery used in the garment textile industry, the machines have been adapted to deposit and stitch fiber roving material onto a base material. Roving material, mostly common carbon fibers, from about 3.000 up to 50.000 filaments can be applied. The preform is produced continuously by the placement of a single roving. The roving material pulled of a spool is guided by a pipe which is positioned in front of the stitching needle. The roving pipe and the frame, where the base material is fixed onto, move synchronized stepwise to perform a zigzag stitch relative to needle position. The stitching head equipped with roving spool, pipe and needle can rotate

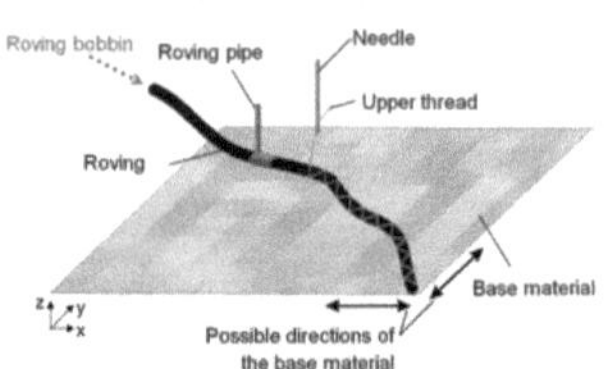

Sketch of the princple of the tailored fiber placement manufacturing process

arbitrarily 360 degrees. During each stitch the upper threat is pulled through the base material and looped around the lower threat spool. Hence a double backstitch is performed. Currently, up to 800 stitches per minute can be achieved. The base material can be a 2D-textile such as woven or non-woven fabric or a matrix-compatible foil material for thermoplastic composites. The stitching path can be designed in form of a pattern either with the help of classical design embroidery software or more recent by use of 2D-CAD systems. Afterwards necessary information of the stitch positions are added to the pattern with the help of so called punch software and finally transferred to the TFP machine.

The infiltration of TFP-preforms can be done with conventional processing techniques such as resintransfer moulding, vacuum bag molding, pressing and autoclave moulding. In the case of thermoplastic composites the matrix material and the reinforcement fibers can be placed simultaneously e.g. in the form of films or fibers. The base material can then be a thermoplastic foil which melts during the consolidation process and becomes part of the matrix. This type is ideally suited for deep-drawn TFP-preforms.

Advantages of the TFP technology

- Net-shape manufacturing reduces costs and waste of valuable reinforcement fibers, e.g. carbon fibers

- Automatic deposition ensures high accuracy and repeatability of amount and orientation of fibers

- TFP machines with multiple heads can be applied to achieve a reasonable productivity, each head is manufacturing in a synchronized way the same preform

- Fibers can be orientated in arbitrary direction in order to manufacture highly stress adapted composite parts

- A variety of fibers such as carbon, glass, basalt, aramid, natural, thermoplastic, ceramic fibers and also metallic threads can be applied and combined within one perform

Applications for structural parts

The TFP technology allows the fabrication of preforms tailored for specific composite components or reinforcements. Applications range from highly accelerated lightweight parts for industrial robots or blades for compressors up to CFRP aircraft parts, e.g. I-beam for the NH-90 helicopter, automotive structures and bicycle parts.[2]

TFP manufactured prefoms made of carbon and glass fibers for structural FRP parts

TFP for self-heating tooling and components

Using the carbon roving as an electric heating element offers the possibility to manufacture composite structures with embedded heating layers. Due to the high flexibility in the design of the heating pattern an overall nearly equal heat distribution can be achieved. In terms of applications this technology embedded in solid composite molds is very beneficial for resin consolidation and binder activation in out-of-autoclave processes. Composite molds show similar heat expansion properties as the manufactured composite parts. The lower

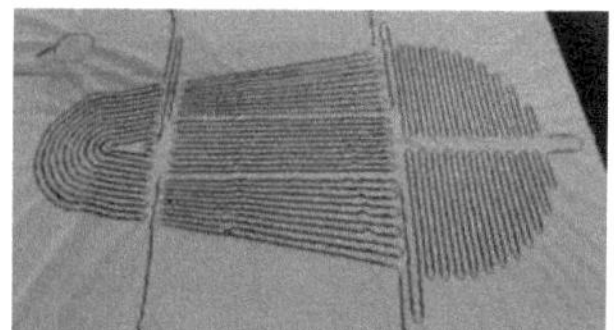

Preform of a carbon layer heating structure

thermal mass of composite tools compared to common metal molds help to shorten the manufacturing cycle of FRP parts and decrease the energy need for the production process. Further the TFP heating elements can be applied in CFRP wing structures of airplanes or blades of wind mills for anti- and de-icing tasks. The TFP structure embedded in elastomeric heating bags can applied to manufacturing or repairing processes of composite parts.[3]

References

[1] Journal of Reinforced Plastics and Composites June 1998 vol. 17 no. 9: "Tailored Fiber Placement-Mechanical Properties and Applications" (http://jrp.sagepub.com/content/17/9/774.short)

[2] Hightex Verstärkungsstrukturen GmbH (http://www.hightex-dresden.de/Das-Unternehmen/Von-der-Idee-zum-Unternehmen/44511/)

[3] Qpoint Composite GmbH (http://www.qpoint-composite.de/Produkte/44111/)

External links

- EU-Project "Embroidery" (http://www.embroidery-project.eu)
- Tailored Fiber Placement at the IPF Dresden (http://www.ipfdd.de/tfp-technology)

Composite material

Composite materials, often shortened to **composites** or called **composition materials**, are engineered or naturally occurring materials made from two or more constituent materials with significantly different physical or chemical properties which remain separate and distinct at the macroscopic or microscopic scale within the finished structure.

A common example of a composite would be disc brake pads, which consists of hard ceramic particles embedded in soft metal matrix. Another example is found in shower stalls and bathtubs which are made of fibreglass. Imitation granite and cultured marble sinks and countertops are also widely used. The most advanced examples perform routinely on spacecraft in demanding environments.

A cloth of woven carbon fiber filaments, a common element in composite materials

Wattle and daub is one of the oldest manmade composite materials, at over 6000 years old.[1] Concrete is also a composite material, and is used more than any other man-made material in the world.[2] As of 2006, about 7.5 billion cubic metres of concrete are made each year—more than one cubic metre for every person on Earth.[3]

Composition

Wood is a natural composite of Cellulose fibres in a matrix of lignin.[4][5] The earliest man-made composite materials were straw and mud combined to form bricks for building construction. The ancient brick-making process can still be seen on Egyptian tomb paintings in the Metropolitan Museum of Art.

Plywood is a commonly encountered composite material.

Composites are made up of individual materials referred to as constituent materials. There are two categories of constituent materials: matrix and reinforcement. At least one portion of each type is required. The matrix material surrounds and supports the reinforcement materials by maintaining their relative positions. The reinforcements impart their special mechanical and physical properties to enhance the matrix properties. A synergism produces material properties unavailable from the individual constituent materials, while the wide variety of matrix and strengthening materials allows the designer of the product or structure to choose an optimum combination.

Engineered composite materials must be formed to shape. The matrix material can be introduced to the reinforcement before or after the reinforcement material is placed into the mould cavity or onto the mould surface. The matrix material experiences a melding event, after which the part shape is essentially set. Depending upon the nature of the matrix material, this melding event can occur in various ways such as chemical polymerization or solidification from the melted state.

A variety of moulding methods can be used according to the end-item design requirements. The principal factors impacting the methodology are the natures of the chosen matrix and reinforcement materials. Another important factor is the gross quantity of material to be produced. Large quantities can be used to justify high capital expenditures for rapid and automated manufacturing technology. Small production quantities are accommodated with lower capital expenditures but higher labour and tooling costs at a correspondingly slower rate.

Most commercially produced composites use a polymer matrix material often called a resin solution. There are many different polymers available depending upon the starting raw ingredients. There are several broad categories, each with numerous variations. The most common are known as polyester, vinyl ester, epoxy, phenolic, polyimide, polyamide, polypropylene, PEEK, and others. The reinforcement materials are often fibres but also commonly ground minerals. The various methods described below have been developed to reduce the resin content of the final product, or the fibre content is increased. As a rule of thumb, lay up results in a product containing 60% resin and 40% fibre, whereas vacuum infusion gives a final product with 40% resin and 60% fibre content. The strength of the product is greatly dependent on this ratio.

Moulding methods

In general, the reinforcing and matrix materials are combined, compacted and processed to undergo a melding event. After the melding event, the part shape is essentially set, although it can deform under certain process conditions. For a thermoset polymeric matrix material, the melding event is a curing reaction that is initiated by the application of additional heat or chemical reactivity such as an organic peroxide. For a thermoplastic polymeric matrix material, the melding event is a solidification from the melted state. For a metal matrix material such as titanium foil, the melding event is a fusing at high pressure and a temperature near the melt point.

For many moulding methods, it is convenient to refer to one mould piece as a "lower" mould and another mould piece as an "upper" mould. Lower and upper refer to the different faces of the moulded panel, not the mould's configuration in space. In this convention, there is always a lower mould, and sometimes an upper mould. Part construction begins by applying materials to the lower mould. Lower mould and upper mould are more generalized descriptors than more common and specific terms such as male side, female side, a-side, b-side, tool side, bowl, hat, mandrel, etc. Continuous manufacturing processes use a different nomenclature.

The moulded product is often referred to as a panel. For certain geometries and material combinations, it can be referred to as a casting. For certain continuous processes, it can be referred to as a profile. Applied with a pressure roller, a spray device or manually. This process is generally done at ambient temperature and atmospheric pressure. Two variations of open moulding are Hand Layup and Spray-up.

Vacuum bag moulding

A process using a two-sided mould set that shapes both surfaces of the panel. On the lower side is a rigid mould and on the upper side is a flexible membrane or vacuum bag. The flexible membrane can be a reusable silicone material or an extruded polymer film. Then, vacuum is applied to the mould cavity. This process can be performed at either ambient or elevated temperature with ambient atmospheric pressure acting upon the vacuum bag. Most economical way is using a venturi vacuum and air compressor or a vacuum pump.

A vacuum bag is a bag made of strong rubber-coated fabric or a polymer film used to bond or laminate materials. In some applications the bag encloses the entire material, or in other applications a mold is used to form one face of the laminate with the bag being single sided to seal the outer face of the laminate to the mold. The open end is sealed and the air is drawn out of the bag through a nipple using a vacuum pump. As a result, uniform pressure approaching one atmosphere is applied to the surfaces of the object inside the bag, holding parts together while the adhesive cures. The entire bag may be placed in a temperature-controlled oven, oil bath or water bath and gently heated to accelerate curing.

In commercial woodworking facilities vacuum bags are used to laminate curved and irregular shaped workpieces.

Vacuum bagging is widely used in the composites industry as well. Carbon fiber fabric and fiberglass, along with resins and epoxies are common materials laminated together with a vacuum bag operation.

Typically, polyurethane or vinyl materials are used to make the bag, which is commonly open at both ends. This gives access to the piece, or pieces to be glued. A plastic rod is laid onto the bag, which is then folded over the rod. A plastic sleeve with an opening in it, is then snapped over the rod. This procedure forms a seal at both ends of the bag, when the vacuum is ready to be drawn.

A "platen" is used inside the bag for the piece being glued to lay on. The platen has a series of small slots cut into it, to allow the air under it to be evacuated. The platen must have rounded edges and corners to prevent the vacuum from tearing the bag.

When a curved part is to be glued in a vacuum bag, it is important that the pieces being glued be placed over a solidly built form, or have an air bladder placed under the form. This air bladder has access to "free air" outside the bag. It is used to create an equal pressure under the form, preventing it from being crushed.[6]

Pressure bag moulding

This process is related to vacuum bag moulding in exactly the same way as it sounds. A solid female mould is used along with a flexible male mould. The reinforcement is placed inside the female mould with just enough resin to allow the fabric to stick in place (wet lay up). A measured amount of resin is then liberally brushed indiscriminately into the mould and the mould is then clamped to a machine that contains the male flexible mould. The flexible male membrane is then inflated with heated compressed air or possibly steam. The female mould can also be heated. Excess resin is forced out along with trapped air. This process is extensively used in the production of composite helmets due to the lower cost of unskilled labor. Cycle times for a helmet bag moulding machine vary from 20 to 45 minutes, but the finished shells require no further curing if the moulds are heated.

Autoclave moulding

A process using a two-sided mould set that forms both surfaces of the panel. On the lower side is a rigid mould and on the upper side is a flexible membrane made from silicone or an extruded polymer film such as nylon. Reinforcement materials can be placed manually or robotically. They include continuous fibre forms fashioned into textile constructions. Most often, they are pre-impregnated with the resin in the form of prepreg fabrics or unidirectional tapes. In some instances, a resin film is placed upon the lower mould and dry reinforcement is placed above. The upper mould is installed and vacuum is applied to the mould cavity. The assembly is placed into an autoclave. This process is generally performed at both elevated pressure and elevated temperature. The use of elevated pressure facilitates a high fibre volume fraction and low void content for maximum structural efficiency.

Resin transfer moulding (RTM)

A process using a two-sided mould set that forms both surfaces of the panel. The lower side is a rigid mould. The upper side can be a rigid or flexible mould. Flexible moulds can be made from composite materials, silicone or extruded polymer films such as nylon. The two sides fit together to produce a mould cavity. The distinguishing feature of resin transfer moulding is that the reinforcement materials are placed into this cavity and the mould set is closed prior to the introduction of matrix material. Resin transfer moulding includes numerous varieties which differ in the mechanics of how the resin is introduced to the reinforcement in the mould cavity. These variations include everything from vacuum infusion (for resin infusion see also boat building) to vacuum assisted resin transfer moulding (VARTM). This process can be performed at either ambient or elevated temperature.

Other

Other types of moulding include press moulding, transfer moulding, pultrusion moulding, filament winding, casting, centrifugal casting and continuous casting. There are also forming capabilities including CNC filament winding, vacuum infusion, wet lay-up, compression moulding, and thermoplastic moulding, to name a few. The use of curing ovens and paint booths is also needed for some projects.[7]

Tooling

Some types of tooling materials used in the manufacturing of composites structures include invar, steel, aluminium, reinforced silicone rubber, nickel, and carbon fiber. Selection of the tooling material is typically based on, but not limited to, the coefficient of thermal expansion, expected number of cycles, end item tolerance, desired or required surface condition, method of cure, glass transition temperature of the material being moulded, moulding method, matrix, cost and a variety of other considerations.

Properties

Mechanics

The physical properties of composite materials are generally not isotropic (independent of direction of applied force) in nature, but rather are typically anisotropic (different depending on the direction of the applied force or load). For instance, the stiffness of a composite panel will often depend upon the orientation of the applied forces and/or moments. Panel stiffness is also dependent on the design of the panel. For instance, the fibre reinforcement and matrix used, the method of panel build, thermoset versus thermoplastic, type of weave, and orientation of fibre axis to the primary force.

In contrast, isotropic materials (for example, aluminium or steel), in standard wrought forms, typically have the same stiffness regardless of the directional orientation of the applied forces and/or moments.

The relationship between forces/moments and strains/curvatures for an isotropic material can be described with the following material properties: Young's Modulus, the shear Modulus and the Poisson's ratio, in relatively simple mathematical relationships. For the anisotropic material, it requires the mathematics of a second order tensor and up to 21 material property constants. For the special case of orthogonal isotropy, there are three different material property constants for each of Young's Modulus, Shear Modulus and Poisson's ratio—a total of 9 constants to describe the relationship between forces/moments and strains/curvatures.

Techniques that take advantage of the anisotropic properties of the materials include mortise and tenon joints (in natural composites such as wood) and Pi Joints in synthetic composites.

Resins

Typically, most common composite materials, including fiberglass, carbon fiber, and Kevlar, include at least two parts, the substrate and the resin.

Polyester resin tends to have yellowish tint, and is suitable for most backyard projects. Its weaknesses are that it is UV sensitive and can tend to degrade over time, and thus generally is also coated to help preserve it. It is often used in the making of surfboards and for marine applications. Its hardener is a MEKP, and is mixed at 14 drops per oz. MEKP is composed of methyl ethyl ketone peroxide, a catalyst. When MEKP is mixed with the resin, the resulting chemical reaction causes heat to build up and cure or harden the resin.

Vinylester resin tends to have a purplish to bluish to greenish tint. This resin has lower viscosity than polyester resin, and is more transparent. This resin is often billed as being fuel resistant, but will melt in contact with gasoline. This resin tends to be more resistant over time to degradation than polyester resin, and is more flexible. It uses the same hardener as polyester resin (at the same mix ratio) and the cost is approximately the same.

Epoxy resin is almost totally transparent when cured. In the aerospace industry, epoxy is used as a structural matrix material or as a structural glue.

Shape memory polymer (SMP) resins have varying visual characteristics depending on their formulation. These resins may be epoxy-based, which can be used for auto body and outdoor equipment repairs; cyanate-ester-based, which are used in space applications; and acrylate-based, which can be used in very cold temperature applications, such as for sensors that indicate whether perishable goods have warmed above a certain maximum temperature.[8] These resins are unique in that their shape can be repeatedly changed by heating above their glass transition temperature (T_g). When heated, they become flexible and elastic, allowing for easy configuration. Once they are cooled, they will maintain their new shape. The resins will return to their original shapes when they are reheated above their T_g.[9] The advantage of shape memory polymer resins is that they can be shaped and reshaped repeatedly without losing their material properties, and these resins can be used in fabricating shape memory composites.[10]

Categories of fiber-reinforced composite materials

Fiber-reinforced composite materials can be divided into two main categories normally referred to as short fiber-reinforced materials and continuous fiber-reinforced materials. Continuous reinforced materials will often constitute a layered or laminated structure. The woven and continuous fibre styles are typically available in a variety of forms, being pre-impregnated with the given matrix (resin), dry, uni-directional tapes of various widths, plain weave, harness satins, braided, and stitched.

The short and long fibers are typically employed in compression moulding

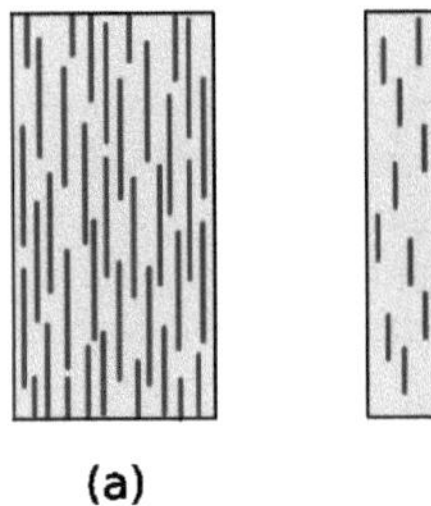
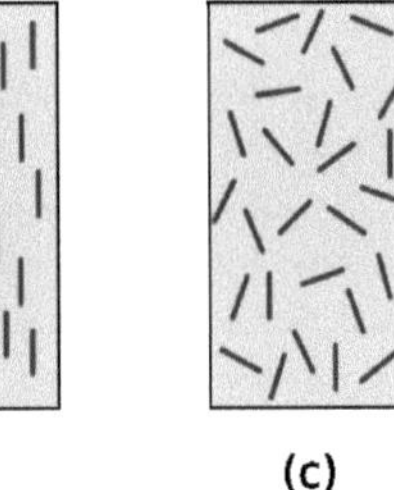

(a)   **(b)**   **(c)**

Typologies of fibre-reinforced composite materials:
a) continuous fibre-reinforced
b) discontinuous aligned fibre-reinforced
c) discontinuous random-oriented fibre-reinforced.

and sheet moulding operations. These come in the form of flakes, chips, and random mate (which can also be made from a continuous fibre laid in random fashion until the desired thickness of the ply / laminate is achieved).

Failure

Shock, impact, or repeated cyclic stresses can cause the laminate to separate at the interface between two layers, a condition known as delamination. Individual fibres can separate from the matrix e.g. fibre pull-out.

Composites can fail on the microscopic or macroscopic scale. Compression failures can occur at both the macro scale or at each individual reinforcing fibre in compression buckling. Tension failures can be net section failures of the part or degradation of the composite at a microscopic scale where one or more of the layers in the composite fail in tension of the matrix or failure the bond between the matrix and fibres.

Some composites are brittle and have little reserve strength beyond the initial onset of failure while others may have large deformations and have reserve energy absorbing capacity past the onset of damage. The variations in fibres and matrices that are available and the mixtures that can be made with blends leave a very broad range of properties that can be designed into a composite structure. The best known failure of a brittle ceramic matrix composite occurred when the carbon-carbon composite tile on the leading edge of the wing of the Space Shuttle Columbia fractured when impacted during take-off. It led to catastrophic break-up of the vehicle when it re-entered the Earth's atmosphere on 1 February 2003.

Compared to metals, composites have relatively poor bearing strength.

Testing

To aid in predicting and preventing failures, composites are tested before and after construction. Pre-construction testing may use finite element analysis (FEA) for ply-by-ply analysis of curved surfaces and predicting wrinkling, crimping and dimpling of composites.[11] Materials may be tested after construction through several nondestructive methods including ultrasonics, thermography, shearography and X-ray radiography[12]

Examples

Materials

Fibre-reinforced polymers or FRPs include wood (comprising cellulose fibres in a lignin and hemicellulose matrix), carbon-fibre reinforced plastic or CFRP, and glass-reinforced plastic or GRP. If classified by matrix then there are thermoplastic composites, short fibre thermoplastics, long fibre thermoplastics or long fibre-reinforced thermoplastics. There are numerous thermoset composites, but advanced systems usually incorporate aramid fibre and carbon fibre in an epoxy resin matrix.

Shape memory polymer composites are high-performance composites, formulated using fibre or fabric reinforcement and shape memory polymer resin as the matrix. Since a shape memory polymer resin is used as the matrix, these composites have the ability to be easily manipulated into various configurations when they are heated above their activation temperatures and will exhibit high strength and stiffness at lower temperatures. They can also be reheated and reshaped repeatedly without losing their material properties. These composites are ideal for applications such as lightweight, rigid, deployable structures; rapid manufacturing; and dynamic reinforcement.[13]

Concrete is probably the most common artificial composite material of all and typically consists of loose stones (aggregate) held with a matrix of cement. Concrete is a very robust material, much more robust than cement, however concrete cannot survive tensile loading. Therefore metal cables are often added to tension the concrete to form reinforced concrete.

Composites can also use metal fibres reinforcing other metals, as in metal matrix composites or MMC. The benefit of magnesium is that it does not degrade in outer space. Ceramic matrix composites include bone (hydroxyapatite reinforced with collagen fibres), Cermet (ceramic and metal) and concrete. Ceramic matrix composites are built primarily for fracture toughness, not for strength. Organic matrix/ceramic aggregate composites include asphalt concrete, mastic asphalt, mastic roller hybrid, dental composite, syntactic foam and mother of pearl. Chobham armour is a special type of composite armour used in military applications.

Additionally, thermoplastic composite materials can be formulated with specific metal powders resulting in materials with a density range from 2 g/cm³ to 11 g/cm³ (same density as lead). The most common name for this type of material is High Gravity Compound (HGC), although Lead Replacement is also used.[14] These materials can be used in place of traditional materials such as aluminium, stainless steel, brass, bronze, copper, lead, and even tungsten in weighting, balancing (for example, modifying the centre of gravity of a tennis racquet), vibration damping, and radiation shielding applications. High density composites are an economically viable option when certain materials are deemed hazardous and are banned (such as lead) or when secondary operations costs (such as machining, finishing, or coating) are a factor.

Engineered wood includes a wide variety of different products such as wood fibre board, plywood, oriented strand board, wood plastic composite (recycled wood fibre in polyethylene matrix), Pykrete (sawdust in ice matrix), Plastic-impregnated or laminated paper or textiles, Arborite, Formica (plastic) and Micarta. Other engineered laminate composites, such as Mallite, use a central core of end grain balsa wood, bonded to surface skins of light alloy or GRP. These generate low-weight, high rigidity materials.

Products

Fiber-reinforced composite materials have gained popularity (despite their generally high cost) in high-performance products that need to be lightweight, yet strong enough to take harsh loading conditions such as aerospace components (tails, wings, fuselages, propellers), boat and scull hulls, bicycle frames and racing car bodies.[15] Other uses include fishing rods, storage tanks, and baseball bats. The new Boeing 787 structure including the wings and fuselage is composed largely of composites. Composite materials are also becoming more common in the realm of orthopedic surgery.

Carbon composite is a key material in today's launch vehicles and heat shields for the re-entry phase of spacecraft. It is widely used in solar panel substrates, antenna reflectors and yokes of spacecraft. It is also used in payload adapters, inter-stage structures and heat shields of launch vehicles. Furthermore disk brake systems of airplanes and racing cars are using carbon/carbon material, and the composite material with carbon fibers and silicon carbide matrix has been introduced in luxury vehicles and sports cars.

In 2007, an all-composite military Humvee was introduced by TPI Composites Inc and Armor Holdings Inc, the first all-composite military vehicle. By using composites the vehicle is lighter, allowing higher payloads. In 2008, carbon fiber and DuPont Kevlar (five times stronger than steel) were combined with enhanced thermoset resins to make military transit cases by ECS Composites creating 30-percent lighter cases with high strength.

Many composite layup designs also include a co-curing or post-curing of the prepreg with various other mediums, such as honeycomb or foam. This is commonly called a sandwich structure. This is a more common layup process for the manufacture of radomes, doors, cowlings, or non-structural parts.[16]

The finishing of the composite parts is also critical in the final design. Many of these finishes will include rain-erosion coatings or polyurethane coatings.

See also

- Aluminium composite panel
- American Composites Manufacturers Association
- Chemical vapour infiltration
- Epoxy granite
- Nanocomposites
- Hybrid material

References

[1] Shaffer, G.D. "An Archaeomagnetic Study of a Wattle and Daub Building Collapse." *Journal of Field Archaeology*, **20**, No. 1. Spring, 1993. 59-75. JSTOR. Accessed 28 January 2007 (http://jstor.org/search)

[2] Lomborg, Bjørn (2001). *The Skeptical Environmentalist: Measuring the Real State of the World.* p. 138. ISBN 978-0-521-80447-9.

[3] "Minerals commodity summary – cement – 2007" (http://minerals.usgs.gov/minerals/pubs/commodity/cement/index.html). US United States Geographic Service. 1 June 2007. . Retrieved 16 January 2008.

[4] http://www.ncsu.edu/bioresources/BioRes_02/BioRes_02_4_534_535_Hubbe_L_BioResJ_Editorial_LoveHate.pdf

[5] David Hon and Nobuo Shiraishi, eds. (2001) Wood and cellulose chemistry, 2nd ed. (New York: Marcel Dekker), p. 5 ff.

[6] http://www.prowoodworkingtips.com/Vacuum_Systems_pg_4_-_Vacuum_bags.html

[7] PCT (http://www.pactinc.com/compositematerials.asp)

[8] Environmental Sensors (http://www.crgrp.net/technology/systemsportfolio/environmental-sensors.shtml)

[9] "Shape Memory Polymers: An Overview" (http://www.crgrp.com/technology/overviews/smp1.shtml). Cornerstone Research Group. . Retrieved 2009-09-30.

[10] "Shape Memory Polymer Resins" (http://www.crgrp.com/technology/materialsportfolio/veriflex.shtml). Cornerstone Research Group. . Retrieved 2009-09-30.

[11] Waterman, Pamela J.. "The Life of Composite Materials" (http://66.195.41.10/Articles/Feature/The-Life-of-Composite-Materials-200704101800.html). *Desktop Engineering Magazine* **April 2007**. .

[12] Matzkanin, George A.; Yolken, H. Thomas. "Techniques for the Nondestructive Evaluation of Polymer Matrix Composites" (http://ammtiac.alionscience.com/pdf/AQV2N4.pdf). *AMMTIAC Quarterly* **2** (4). .

[13] "Shape Memory Composites" (http://www.crgrp.com/technology/overviews/composites.shtml). Cornerstone Research Group. . Retrieved 2009-10-02.

[14] Material Properties Data: High Gravity Compound (HGC) (http://www.makeitfrom.com/data/?material=HGC)

[15] "Rubbn'Repair Composite Repair System" (http://www.rubbnrepair.com/). CRG Industries, LLC. . Retrieved 2009-10-02.

[16] Vantage Composites and Thermoforming, Inc. http://www.vantagecandt.com

Further reading

- Autar K. Kaw (2005). *Mechanics of Composite Materials* (2nd ed.). CRC. ISBN 0-84-931343-0.
- Handbook of Polymer Composites for Engineers By Leonard Hollaway Published 1994 Woodhead Publishing
- Matthews, F.L. & Rawlings, R.D. (1999). *Composite Materials: Engineering and Science.* Boca Raton: CRC Press. ISBN 0-84-930621-3.

External links

- Composites UK - The official UK trade association with a library of online information about composite materials (http://www.compositesuk.co.uk)
- Composite material key concepts (http://www.science.org.au/nova/059/059key.htm)
- Distance learning course in polymers and composites (http://www3.open.ac.uk/courses/bin/p12. dll?C01T838)
- Composite Sandwich Structure of Minardi F1 Car (http://www.nenastran.com/newnoran/chPDF/CASE_Chassis_Design.pdf)
- Teaching support materials for the University of Plymouth composites degree (http://www.tech.plym.ac.uk/sme/mats324/)

3D composites

3D composites are a way to increase through-the-thickness mechanical properties of laminates by having reinforcement in the thickness direction (out of plane). When compared to a 2D composite, the impact resistance, compression after impact (CAI), and delamination control is significantly improved without significantly reducing the mechanical properties along the plane. Possible applications include wind turbine blades, ballistic armor, boat construction, and the automotive industry. The most common manufacturing technique is resin transfer molding.

3T Cycling

3T is an Italian cycle sport company associated with many champion cyclists. It was founded in 1961 and soon won a reputation for lightweight racing cycle componentry.

It pioneered the use of aerospace-grade aluminum alloys, and worked closely with pro cyclists to shape handlebars to their exact requirements. 3T's first aero bar set the pattern for all cyclists aiming for sheer speed against the clock.

3T switched production to carbon-fiber composite materials and in 2008 returned to pro cycling after several years' absence. For the 2008 season it sponsored the CSC team, which won the Tour de France. 3T sponsors three pro teams for the 2009 professional season.

The quest for light weight

3T is an Italian cycle sport firm, located in Madone, near Milano. It was originally known as 3TTT — Tecnologia del Tubo Torinese (Turin Tube Technology). Many competitions have been won on 3T components, including the Tour de France, Olympic races, World Championships, and the World Hour Record.

The firm was founded by Mario Dedioniggi in Torino in 1961. Dedioniggi was skilled at manipulating and bending steel tubes to fabricate the lightweight handlebars desired by racing cyclists, and Italian professional riders were among 3T's first customers.

By 1970 3T handlebars and stems were in widespread use in the European professional peloton. In the quest for lighter weight, 3T switched production to aluminum alloy in place of steel. It was among the first to use aluminum for these components, where strength is critical to safety.

In 1975 it produced the Superleggera weighing 250 grammes — claimed to be the world's lightest drop handlebar. This bar was the first to be made of 7075 aluminium alloy, a material usually used in aerospace applications. The high strength-to-weight ratio of this alloy, also known as Ergal, found uses in other sports: 3T ski poles sold well for several seasons during the 1970s.

Aerodynamics arrive

The firm worked closely with professional racers to refine the design of their handlebars. The various 'bends' took their name from the champions of the era — Merckx, Saronni, Moser, and Gimondi. In 1984 Francesco Moser used a newly-developed 3T bar to capture the world hour record, breaking through the 50-kilometre barrier for the first time.

This new 'bullhorn' bar put the rider in a lower, more aerodynamic position[1] for greater speed. Most riders in search of pure speed now use a similar position. In Moser's honor, 3T named the bar '51.151' — the distance Moser covered in kilometres (31.784 miles) at the Mexico City velodrome on January 23, 1984.

Transatlantic alliances

3T continued to develop new handlebar designs to assist riders reach their goal of greater speed. The growing power in world cycling, the United States, started to take notice. Scott USA commissioned 3T to create the first handlebars specific to triathlon use, and in a departure from its roots in road racing, 3T partnered with Salsa Cycles to create a range of mountain-bike components. Success was immediate: in 1994, Nicolas Vouilloz won the third of his many World Championship downhill titles in Vail (CO) on a 3T bar.

New technology drivers

Bicycle components underwent radical change in the 1990s. The traditional quill handlebar stem was rendered obsolete by the introduction of the threadless headset. 3T responded with the industry's first forged stem for this application, the ForgeAhead. Other manufacturers followed 3T's lead and switched to this technique, abandoning welded stems.

By the late 1990s, carbon-fiber composites were becoming the major driver of change. 3T, best known for its skill in aluminum alloy fabrication, took up the challenge of re-engineering to build components in composite materials. Race successes continued with Gold in the World Championships in 2000 and 2002 and a string of popular wins for Erik Zabel in the Italian home classic, Milan – San Remo, but it was not until 2008 that 3T again saw victory in a major tour.

Renewed race success

For the 2008 season, 3T sponsored Team CSC. The team debuted a new, all-composites aerofoil time-trial bar, the 3T Ventus. The team's World Champion time trialist, Fabian Cancellara, rode this to victory[2] at the Prologue of the Amgen Tour of California. Moving on to the European season at the Giro d'Italia, CSC's youthful team narrowly failed to win the race's opening Team Time Trial.

Great success followed at the Tour de France.[3] CSC captain Carlos Sastre won the Yellow Jersey[4] of the Overall winner, his team mate Andy Schleck won the White Jersey for Best Young Rider, and CSC won Best Team. Race reports noted Sastre's outstanding performance[5] in limiting his losses in the final weekend's Individual Time Trial, where he was widely expected[6] to lose the lead to his nearest rival, Cadel Evans.

Riding for Switzerland at the Beijing Olympics shortly after, Cancellara won Gold[7] in the Men's Time Trial,[8] beating his team mate Gustav Larsson, riding for Sweden, into second place, both riders using the Ventus bar. Their team captain Carlos Sastre animated the Men's Road Race, while Cancellara eventually placed third,[9] one of three riders in the first five finishers riding 3T.[10]

For the 2009 season, 3T is sponsoring three professional teams[11] : Cervélo TestTeam (captained by Carlos Sastre[12]), Garmin Slipstream, and Milram.

References

[1] http://www.sportmedicina.com/CAMPIONI7/MOSER.jpg Francesco Moser winning the World Hour Record in Mexico City, 1984

[2] http://www.amgentourofcalifornia.com/Archives/2008-archive/stages/prologue.html Fabian Cancellara triumphs at Stanford University

[3] http://www.letour.fr/2008/TDF/LIVE/fr/2100/classement/index.html Tour de France final classifications

[4] http://tour-de-france.velonews.com/photo/80845 *VeloNews* pictures Carlos Sastre in full flight

[5] http://autobus.cyclingnews.com/road/2008//tour08/?id=results/tour0820 *Cycling News*: Sastre rides the time trial of his life

[6] http://www.velonews.com/article/80660 *VeloNews*: Johan Bruyneel tips Evans for the overall win

[7] http://results.beijing2008.cn/WRM/ENG/INF/CR/C73T/CRM011101.shtml#CRM011101 Fabian Cancellara wins the Gold Medal in the Men's Olympic Time Trial

[8] http://www.velonews.com/photo/81631 *VeloNews*: Cancellara becomes the king of the time trial

[9] http://results.beijing2008.cn/WRM/ENG/INF/CR/C73R/CRM012101.shtml#CRM012101 Bronze for Cancellara in the Olympic Road Race

[10] http://autobus.cyclingnews.com/photos/2008/olympics08/index.php?id=/photos/2008/olympics08/1/DV372900 The top five finishers duke it out in Beijing

[11] http://www.thenew3t.com/content.aspx?m=About&i=PressKit 3T US press communique, Interbike 2008

[12] http://www.velonews.com/article/82795 *VeloNews*: 'It's official: Sastre to Cervélo'

Fiber cement siding

Fiber cement siding (or "fibre cement cladding" in the UK and Australasia) is a building material used to cover the exterior of a building in both commercial and domestic applications.

Usage

Fiber cement is a composite material made of sand, cement and cellulose fibers. In appearance fiber cement siding most often consists of overlapping horizontal boards, imitating wooden siding, clapboard and imitation shingles. Fiber cement siding is also manufactured in a sheet form and is used not only as cladding but is also commonly used as a soffit / eave lining and as a tile underlay on decks and in bathrooms.

Hardipanels on design-build addition, Ithaca NY

Fiber cement siding is not only used as an exterior siding, it can also be utilized as a substitute for timber fascias and barge boards in high fire areas.

Specifications

Sheet sizes vary slightly from manufacturer to manufacturer but generally they range between 2400 – 3000 mm in length and 900 –1200mm in width (600 & 450 mm increments). This manufactured size minimizes on-site wastage as residential floor, wall and roof structures lay structural members at 450 or 600 centres.

Fibre cement thicknesses vary between 4.5-18mm and also vary in density – the lower density resulting in a fibrous rough edge when cut and the higher density having a cleaner smoother edge when cut.

Thermal resistance and sound transmission vary greatly between fiber cement products. Fiber cement sheet products rate poorly in thermal resistance and sound transmission and separate wall insulation is highly recommended. Generally the thicker and denser the product the better resistance it will have to temperature and sound transmission.

CSR Fiber Cement sheet cladding - dwelling addition, Hardys Bay, NSW, Australia

Installation

Fibre cement cladding is a very heavy product and requires two people to carry the uncut sheets. Thin fibre cement cladding is fragile before installation and must be handled carefully; it is prone to chipping and breakage if improperly handled.

Once the product is cut it may again require two people to install – one to hold the sheet flush against studwork and the other to nail the product in place.

Cutting fibre cement cladding sheeting usually requires a mechanised saw or metal hand shears and sheets can be cut to size in three ways:

- Thinner sheets can be scored with a heavy duty cutting blade and snapped
- Purpose made "fibro cutter" (an Australian term)
- Thicker and denser sheets require cutting by a mechanical saw

Some caution must be exercised to properly ventilate areas where fiber cement siding (FCS) is being cut; long-term exposure to the silica dust generated during the installation process can cause silicosis.

Fibre cement cladding can be painted before or after installation. (For areas of exposure, weatherproof paint must be used.) Once the product is fixed the joints are usually covered with timber battens and the entire wall surface is painted.

History

Fibre cement products came about as a replacement for the widely used "Asbestos Cement Sheeting" product manufactured by "James Hardie" until the late 1980s.

Durability

The external cladding products require very little maintenance once installed and painted. The thicker/denser fiber cement products have excellent impact resistance but the thinner less dense products need to be protected from impact. Compared to wooden siding, fiber cement is not susceptible to termites or rot.[1]

Detail - timber battens on fibre cement cladding, dwelling addition, Hardys Bay, NSW, Australia

Fire Resistance

Fibre cement cladding is a non combustible material which is widely used in high bush fire prone areas throughout Australia . Pictured above is James Hardie's Fiber Cement cladding 'Scyon Linea' weatherboard which has been substituted for the traditional timber fascia and barge board materials.

Alternatives

Competitors to fiber cement cladding include products made from vinyl, polyvinyl chloride, wood composite products (such as hardboard and Masonite) and aluminum siding

See also

- Cement board
- Fibro
- Eternit

References

[1] Ball, John E. "Mineral-Fiber Siding". *Light construction techniques: from foundation to finish*. Reston, VA: Reston. p. 189. ISBN 0835940357.

External links

- Installing Fiber Cement Siding (http://www.hammerzone.com/archives/exterior/siding/replace1/ fiber_cement/install.htm)

Fiber pull-out

Fiber pull-out is one of the failure mechanisms in fiber-reinforced composite materials[1] . Other forms of failure include delamination, intralaminar matrix cracking, longitudinal matrix splitting, fiber/matrix debonding, and fiber fracture[1] . The cause of fiber pull-out and delamination is weak bonding[2] .

Work for debonding, [3]

where

- is fiber diameter
- is failure strength of the fiber
- is the length of the debonded zone
- is fiber modulus

In ceramic matrix composite material this mechanism is not a failure mechanism, but essential for its fracture toughness, which is several factors above that of conventional ceramics.

The figure is an example for how a fracture surface of this material looks like. The strong fibers form bridges over the cracks before they fail at elongations around 0.7%, and thus prevent brittle rupture of the material at 0.05%, especially under thermal shock conditions. [4] This allows using this type of ceramics for heat shields applied for the re-entry of space vehicles, for disk brakes and slide bearing components.

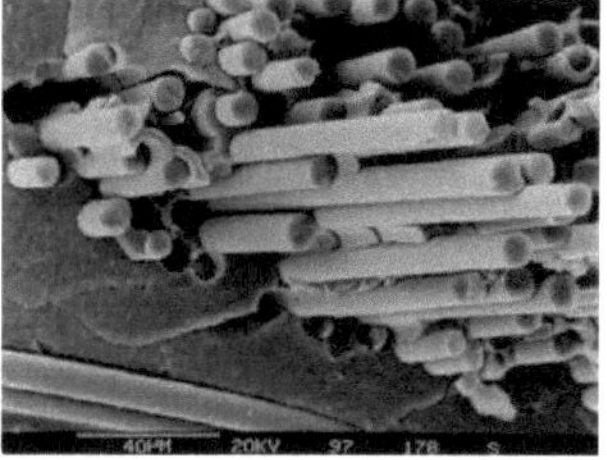

References

[1] WJ Cantwell, J Morton (1991). "The impact resistance of composite materials -- a review". *Composites* **22** (5): 347–62. doi:10.1016/0010-4361(91)90549-V.

[2] Serope Kalpakjian, Steven R Schmid. "Manufacturing Engineering and Technology". International edition. 4th Ed. Prentice Hall, Inc. 2001. ISBN 0-13-017440-8

[3] PWR Beaumont. "Fracture mechanisms in fibrous composites". Fracture Mechanics, Current Status, Future Prospects. Edited by RA Smith. Pergamon Press: 1979. p211-33 in WJ Cantwell, J Morton (1991). "The impact resistance of composite materials -- a review". *Composites* **22** (5): 347–62. doi:10.1016/0010-4361(91)90549-V.

[4] W. Krenkel, ed.:*Ceramic Matrix Composites*, Wiley-VCH, Weinheim, 2008, ISBN 978-3-527-31361-7

Fiber-reinforced composite

A **fiber-reinforced composite** (FRC) consists of three components: (i) the fibers as the discontinuous or dispersed phase, (ii) the matrix as the continuous phase, and (iii) the fine interphase region, also known as the interface[1] [2]. This is a type of advanced composite group, which makes use of rice husk, rice hull, and plastic as ingredients. This technology involves a method of refining, blending, and compounding natural fibers from cellulosic waste streams to form a high-strength fiber composite material in a polymer matrix. The designated waste or base raw materials used in this instance are those of waste thermoplastics and various categories of cellulosic waste including rice husk and saw dust.

FRC is high-performance fiber composite achieved and made possible by cross-linking cellulosic fiber molecules with resins in the FRC material matrix through a proprietary molecular re-engineering process, yielding a product of exceptional structural properties.

Through this feat of molecular re-engineering selected physical and structural properties of wood are successfully cloned and vested in the FRC product, in addition to other critical attributes to yield performance properties superior to contemporary wood.

This material, unlike other composites, can be recycled up to 20 times, allowing scrap FRC to be reused again and again.

The failure mechanisms in FRC materials include delamination, intralaminar matrix cracking, longitudinal matrix splitting, fiber/matrix debonding, fiber pull-out, and fiber fracture [1].

Fiber-reinforced composite

Difference between wood plastic composite and fiber-reinforced composite:

Features	Plastic lumber	Wood plastic composite	FRC	Wood
Recyclable	Yes	No	Yes	No
House Construction	No	No	Yes	Yes
Water Absorption	0.00%	0.8% and above	0.3% and below	10% and above

Properties

Tensile Strength	ASTM D 638	15.9 MPa
Flexural Strength	ASTM D 790	280 MPa
Flexural Modulus	ASTM D 790	1582 MPa
Failure Load	ASTM D 1761	1.5 KN - 20.8 KN
Compressive Strength		20.7MPa
Heat Reversion	BS EN 743 : 1995	0.45%
Water Absorption	ASTM D 570	0.34%
Termite Resistant	FRIM Test Method	3.6

Application

There are also applications in the market, which utilize only waste materials. Its most widespread use is in outdoor deck floors, but it is also used for railings, fences, landscaping timbers, cladding and siding, park benches, molding and trim, window and door frames, and indoor furniture. See for example the work of *Waste for Life*, which collaborates with garbage scavenging cooperatives to create fiber-reinforced building materials and domestic problems from the waste their members collect: Homepage of Waste for Life [3]

See also

* Fracture mechanics
* Plastic lumber
* Wood plastic composite
* Fibre-reinforced plastic

References

[1] WJ Cantwell, J Morton (1991). "The impact resistance of composite materials -- a review". *Composites* **22** (5): 347–62. doi:10.1016/0010-4361(91)90549-V.

[2] Serope Kalpakjian, Steven R Schmid. "Manufacturing Engineering and Technology". International edition. 4th Ed. Prentice Hall, Inc. 2001. ISBN 0-13-017440-8.

[3] http://wasteforlife.org/

Nu-Wood Decorative Millwork

Nu-Wood Decorative Millwork is millwork made of a specially formulated polyurethane polymer. The standard density is similar to white pine. It can be cut, sawn, shaped, routed, nailed, stapled, and screwed just like wood. Profiles exist for a variety of uses, indoors and out. Advantages over wood include a surface which will not crack, flake, or blister and a product which is resistant to water, mold, mildew, and insects.

Manufacturing process

The polyurethane foam is created from two components which generate heat when combined. One component contains water and the other component reacts with water. The oxygen of the water fuels the exothermic reaction, leaving hydrogen bubbles which causes the compound to foam. Given no compression, the foam rises to fill the space allowed. Based on the mixture and the specific makeup of the two chemicals, it is possible to estimate exactly how much foam is required to fill a given space. More importantly, it is possible to determine the exact amount of mixture required to fill a defined space and achieve a specific packing ratio.

Tremendous compression is required to obtain packing equal to the density of white pine. In addition, fake wood needs a wood look and feel on the surface. This means the finished product must easily slide out of the mold that provides its shape. It also means the mold must be strong enough to withstand the pressure of the expanding foam. This combination of requirements creates a complex problem that is solved by a variety of techniques.

Silicone rubber is used to create negative space molds from original millwork created by expert woodworkers. The forms are polished, painted with a release agent surrounded by a box made of angled aluminum and filled with the liquid silicone rubber compound. The rubber is generally around one inch thick. Once the rubber hardens the form is removed from the standard, which is shelved for future use. At this time, the form provides a negative space with the exact dimensional and surface characteristics of the woodwork to be produced, plus a small tolerance for expected shrinkage.

The rubber form and form box pass along a bed of rollers to a paint shop where the rubber form is covered with a thin veneer of water-based paint. The heat and pressure of the expanding foam cause the paint to transfer to the finished piece. The paint protects the rubber form from the chemicals in the polyurethane and provides a release from the piece.

While the finished polyurethane product is quite light, consisting largely of closed cells of hydrogen, the rubber form and the metal and wood foam box are very heavy. The manufacturing process is difficult to automate due to the many shapes and sizes and the fact that most orders are unique. Thus, the work requires great strength and is quite hazardous to the back, hands, and feet. Competition produces pricing pressures, and labor is the most expensive component of pricing. As a result, low wages are standard.

History

Nu-Wood Decorative Millwork is manufactured in Goshen, Indiana. The business was originally founded to serve the manufactured housing industry by Robert DuBois, who later sold the company to Goshen Rubber Company. Goshen Rubber sold the company to private investors in 1998. In 2001 Nu-Wood was acquired by the Beil Family who operated Nu-Wood building a strong presence in Indianapolis and Cincinnati. In 2010 Nu-Wood was purchased by Len and Marci Morris who also operate Stairsupplies.com. Nu-Wood operates delivery vehicles and serving the continental US and Canada.

On 6 December 2001, a worker argument resulted in the shooting death of the general manager, the wounding of six employees and the shooter's suicide. A new operations manager joined the company in 1993 and worked to create a more friendly and stress-free work environment. For more information: http://www.cnn.com/2001/US/12/06/indiana.shooting/

External links

- Nu-Wood.com [1]

References

[1] http://www.nu-wood.com/

Fiber volume ratio

Fiber volume ratio is an important mathematical element in composite engineering.

[1]

where:

is the fiber volume ratio

and

is the volume of fibers

is the volume of composite

One of the standardized method of determining the fiber volume ratio is the ignition or burnout method [1], described in ASTM specification D2584-02. The described method can be applied to composites containing inorganic fibers in an organic matrix [1].

Another standardized method is the acid digestion method which is explained in ASTM specification D3171-99. The method attacks the matrix that is acid-soluble but at the same time does not attack the fiber [2].

The fiber volume ratio is used to calculate the *void volume ratio*.

[1]

where:

is the void volume ratio

and

is the matrix volume ratio

is the volume of voids

References

[1] Isaac M Daniel, Ori Ishai. (2006).*Engineering Mechanics of Composite Materials*. 2nd ed. Oxford University Press. ISBN 978-0-19-532244-6

[2] (2004). *Space Simulation; Aerospace and Aircraft; Composite Materials* in *Annual book of ASTM Standards*. Vol 15.03. American Society for Testing and Materials. West Conshohocken. PA.

Fiber-reinforced concrete

Fiber-reinforced concrete (FRC) is concrete containing fibrous material which increases its structural integrity. It contains short discrete fibers that are uniformly distributed and randomly oriented. Fibers include steel fibers, glass fibers, synthetic fibers and natural fibers. Within these different fibers that character of fiber-reinforced concrete changes with varying concretes, fiber materials, geometries, distribution, orientation and densities.

Historical perspective

The concept of using fibers as reinforcement is not new. Fibers have been used as reinforcement since ancient times. Historically, horsehair was used in mortar and straw in mud bricks. In the early 1900s, asbestos fibers were used in concrete, and in the 1950s the concept of composite materials came into being and fiber-reinforced concrete was one of the topics of interest. There was a need to find a replacement for the asbestos used in concrete and other building materials once the health risks associated with the substance were discovered. By the 1960s, steel, glass (GFRC), and synthetic fibers such as polypropylene fibers were used in concrete, and research into new fiber-reinforced concretes continues today.

Effect of fibers in concrete

Fibers are usually used in concrete to control cracking due to both plastic shrinkage and drying shrinkage. They also reduce the permeability of concrete and thus reduce bleeding of water. Some types of fibers produce greater impact, abrasion and shatter resistance in concrete. Generally fibers do not increase the flexural strength of concrete, and so cannot replace moment resisting or structural steel reinforcement. Indeed, some fibers actually reduce the strength of concrete. The amount of fibers added to a concrete mix is expressed as a percentage of the total volume of the composite (concrete and fibers), termed volume fraction (V_f). V_f typically ranges from 0.1 to 3%. Aspect ratio (l/d) is calculated by dividing fiber length (l) by its diameter (d). Fibers with a non-circular cross section use an equivalent diameter for the calculation of aspect ratio. If the modulus of elasticity of the fiber is higher than the matrix (concrete or mortar binder), they help to carry the load by increasing the tensile strength of the material. Increase in the aspect ratio of the fiber usually segments the flexural strength and toughness of the matrix. However, fibers which are too long tend to "ball" in the mix and create workability problems.

Some recent research indicated that using fibers in concrete has limited effect on the impact resistance of the materials[1 & 2]. This finding is very important since traditionally, people think that ductility increases when concrete is reinforced with fibers. The results also indicated out that the use of micro fibers offers better impact resistance compared with the longer fibers.[1]

The High Speed 1 tunnel linings incorporated concrete containing 1 kg/m³ of polypropylene fibers, of diameter 18 & 32 μm, giving the benefits noted below.[1]

Benefits

Polypropylene and Nylon fibers can:

- Improve mix cohesion, improving pumpability over long distances
- Improve freeze-thaw resistance
- Improve resistance to explosive spalling in case of a severe fire
- Improve impact resistance
- Increase resistance to plastic shrinkage during curing

Steel fibers can:

- Improve structural strength
- Reduce steel reinforcement requirements
- Improve ductility
- Reduce crack widths and control the crack widths tightly thus improve durability
- Improve impact & abrasion resistance
- Improve freeze-thaw resistance

Blends of both steel and polymeric fibers are often used in construction projects in order to combine the benefits of both products; structural improvements provided by steel fibers and the resistance to explosive spalling and plastic shrinkage improvements provided by polymeric fibers.

In certain specific circumstances, steel fiber can entirely replace traditional steel reinforcement bar in reinforced concrete. This is most common in industrial flooring but also in some other precasting applications. Typically, these are corroborated with laboratory testing to confirm performance requirements are met. Care should be taken to ensure that local design code requirements are also met which may impose minimum quantities of steel reinforcement within the concrete. There are increasing numbers of tunnelling projects using precast lining segments reinforced only with steel fibers.

Useful standards:

- EN 14889-1:2006 Fibres for Concrete. Steel Fibres. Definitions, specifications & conformity
- EN 14889-2:2006 Fibres for Concrete. Polymer Fibres. Definitions, specifications & conformity
- EN 14845-1:2007 Test methods for fibres in concrete
- ASTM A820-06 Standard Specification for fibres in Fibre Reinforced Concrete
- ASTM C1018-07 Standard test methods for flexural toughness & first crack strength

Some developments in fiber-reinforced concrete

An FRC sub-category named Engineered Cementitious Composite (ECC) claims 500 times more resistance to cracking and 40 percent lighter than traditional concrete. ECC claims it can sustain strain-hardening up to several percent strain, resulting in a material ductility of at least two orders of magnitude higher when compared to normal concrete or standard fiber-reinforced concrete. ECC also claims a unique cracking behavior. When loaded to beyond the elastic range, ECC maintains crack width to below 100 μm, even when deformed to several percent tensile strains. Field results with ECC and The Michigan Department of Transportation resulted in early-age cracking [2].

Recent studies performed on a high-performance fiber-reinforced concrete in a bridge deck found that adding fibers provided residual strength and controlled cracking [3]. There were fewer and narrower cracks in the FRC even though the FRC had more shrinkage than the control. Residual strength is directly proportional to the fiber content.

A new kind of natural fiber-reinforced concrete (NFRC) made of cellulose fibers processed from genetically modified slash pine trees is giving good results. The cellulose fibers are longer and greater in diameter than other timber sources. Some studies were performed using waste carpet fibers in concrete as an environmentally friendly use of recycled carpet waste [4]. A carpet typically consists of two layers of backing (usually fabric from polypropylene tape yarns), joined by $CaCO_3$ filled styrene-butadiene latex rubber (SBR), and face fibers (majority being nylon 6 and nylon 66 textured yarns). Such nylon and polypropylene fibers can be used for concrete reinforcement. Other ideas are emerging to use recycled materials as fibers [5].

For statistical calculations there is a new modelling in the book: B.Wietek, *Stahlfaserbeton*, edited by Vieweg + Teubner, 2008, ISBN 978-3-8348-0592-8.

See also

- Reinforced concrete
- Glass-reinforced plastic
- Fiber-reinforced plastic

References

[1] (http://www.adfil.co.uk/docs/templates/news.asp?monthid=6&yearid=2004)

[2] Li, V., Yang E., Li, M. Field Demonstration of Durable Link Slabs for Jointless Bridge Decks Based on Srtain-Hardening Cementitious Composites – Phase 3: Shrinkage Control, Michigan Department of Transportation, January 2008. www.michigan.gov/.../MDOT_Research_Report_RC-1506_232611_7.pdf

[3] American Concrete Institute 544.3R-93

[4] Concrete Reinforcement with Recycled Fibers by Wang, Y., Wu, HC., and Li, V. Journal of Materials in Civil Engineering/ Nov 2000

[5] Development of recycled PET fiber and its application as concrete-reinforcing fiber http://www.sciencedirect.com/science/article/pii/ S0958946507000273

Related article and publications

- STEEL FIBER PRODUCTS FOR FLOORING AND SHOTCRETING (http://www.bekaert.com/en/ Product Catalog/Application/Construction/Concrete reinforcement. aspx?Industry={4FF19777-D0F7-413A-B75C-8321C9E85D6F}& ProductCategory={C4F50332-AB47-4095-B56B-96964A7F18EF})
- Steel fibre reinforced Concrete FAQ (http://www.stewols.com/feedback.htm)

Orangeburg pipe

Orangeburg pipe (also known as "fiber conduit") is bitumenized fiber pipe made from layers of wood pulp and pitch pressed together. It was used from the 1860s through the 1970s, when it was replaced by PVC pipe for water delivery and ABS pipe for drain-waste-vent (DWV) applications. The name comes from Orangeburg, New York, the town in which most Orangeburg pipe was manufactured. It was manufactured largely by the Fiber Conduit Company, which changed its name to the Orangeburg Manufacturing Company in 1948.

History

The first known use of fiber pipe was in an experimental water delivery pipe in the Boston area. The pipeline, finished in 1867, measured 1.5 miles in length and was in use through 1927. Bitumenized pipe was not in widespread commercial use until the late 19th century when it was utilized exclusively as electrical conduit.

In 1893, Stephen Bradley, Sr. founded the Fiber Conduit Company in Orangeburg, New York. Bradley's neighboring Union Electric Company electric power plants used exhaust steam from their steam generators to dry the fiber conduit before they were sealed with pitch. In turn, the Fiber Conduit Company's conduits were used to run electrical wiring throughout numerous newly-constructed buildings across the country for the next forty years. Bradley, along with several competitors, laid miles of the fiber electrical conduit in sky-scrapers, such as the Empire State Building.

The early 1900s brought massive expansion of the telephone, telegraph and electrical industries along with subway construction along the eastern seaboard. This expansion in the usage of electrical and telecommunications wiring brought with it a rising demand for fiber conduit, which was being used to contain this wiring within buildings, as well as in subway tunnels. In addition, fiber conduit was increasingly being used to create underground ducting for distributing wires under streets and roads, as well as along railroads.

Fiber was next adopted by the booming oil industry to pump salt waste-water from drilling sites to treatment and disposal areas. This industrial use quickly yielded the insight that while long-lived and incredibly durable in normal draining operations, bitumenized fiber easily ruptured under pressure. During this trial usage by the oil industry, the fiber conduit pipe tested was called "Alkacid" by the Fiber Conduit Co. of Orangeburg, New York. Owing to the aforementioned issues with pressurized usage, the oil industry soon stopped using the fiber "Alkacid" pipe and started using cement-asbestos pipe.

While a variety of companies competed with Fiber Conduit Company, it was by far and away the largest producer of bitumenized fiber conduit piping throughout the early 20th century and demand for fiber conduit only increased during World War II with the need for electrical conduit for use in new airfields and military bases. In 1948, the name of the Fiber Conduit Company was changed to the Orangeburg Manufacturing Company. As World War II ended and gave rise to the post-war housing boom, the demand for cheap housing materials was at an all-time high and available drainage materials were scarce. Orangeburg Manufacturing produced a thicker-walled, sturdier, round version of fiber conduit, selling it as "Orangeburg pipe" for sewer and drain uses.

Usage

Orangeburg pipe was made in inside diameters from 2 inches to 18 inches out of wood pulp sealed with hot pitch. Joints were made in a similar fashion and, due to the materials involved, were able to be sealed without the usage of adhesives. Orangeburg was lightweight, albeit brittle, and soft enough to be cut with a handsaw. Orangeburg was a low cost alternative to metal for sewer lines in particular. Lack of strength causes pipes made of Orangeburg to fail more frequently than pipes made with other materials. The useful life for an Orangeburg pipe is about 50 years under ideal conditions, but has been known to fail in as little as 10 years. It has been taken off the list of acceptable materials by most building codes.

It was observed in early usage that, similar to modern PVC piping, Orangeburg was susceptible to deformation from pressure. Thus, manufacturers urged "bedding" the pipes in sand or pea gravel to prevent rupture in much the same fashion that PVC pipes are bedded today.

References

- Orangeburg on sewerhistory.org [1]
- Sewer History [2]

References

[1] http://www.sewerhistory.org/articles/compon/orangeburg/orangeburg.htm

[2] http://www.sewerhistory.org/grfx/components/pipe-orng1.htm

Article Sources and Contributors

Self Drying Concrete Technology (Flooring) *Source*: http://en.wikipedia.org/w/index.php?title=Self_Drying_Concrete_Technology_%28Flooring%29 *Contributors*: Bearcat, Enric Naval, Lamro, Rich Farmbrough, Tivona schneider, Tony1

Concrete *Source*: http://en.wikipedia.org/w/index.php?title=Concrete *Contributors*: (aeropagitica), .:Ajvol:., 119, 123Hedgehog456, 198.207.223.xxx, 777niceweather777, Acroterion, Adaywijaya, Ahering@cogeco.ca, Ahoerstemeier, Ajkgordon, Akonga, Alaaahmadalshaikh, Alan Liefting, Alansohn, Alcwiki, Almulhim, Amatulic, Amcbride, Amgriffith, AnakngAraw, Anders Torlind, Andrej Šalov, Andrejj, Andrevruas, Andy M. Wang, AndyKaotik, Angr, Anna Frodesiak, AnnaFrance, Antandrus, Anubis90909, Archt67, Ard2r, Argyriou, Arifulhasant, Asfasdfasdf, Astronaut, Atlant, AubreyJohnWestonHarrison, Audriusa, Ayls, AzaToth, B.L.A.Z.E, Bachrach44, Baffclan, BarretBonden, Basar, Bazzargh, Ben414, Benbest, Betacommand, Bevo74, Beyond My Ken, BillFlis, Billbeee, Billy boy 97, Biscuittin, Biwiki394, Blocklayer, Bluezy, BobV01, Bobblewik, Bobo192, Bodhi.peace, Bogolov, Bonadea, Boom Goofs, Bossonova, Boulevardier, Brett38655, Brockert, Brookie, Bryan Derksen, Bsadowski1, Bsherr, Bubba hotep, Bucephalus, Buck Mulligan, Burningjoker, Burntsauce, Bwrs, C-M, CHASEMOON, CJ4, Calabe1992, Calvin ngan, Can't sleep, clown will eat me, CanadianCaesar, Carbuncle, CardinalDan, Carlosgtz, CarstenHermann, CasualObserver'48, Causa sui, Cellofsplinter, Chaka84, Charles01, CharlotteWebb, CheMechanical, ChemGardener, Chenopodiaceous, ChicagoConcrete, Chongkian, Christian75, Chuunen Baka, Cimento.Org, Cireshoe, Civil Engineer III, Cjfishy9, Ckatz, Clawed, Commander Keane, ConcreteCuring, Concreteideas, Concretesociety, ConradPino, Contrafool, Conversion script, Corpx, Correogsk, Corrigann, Corvus cornix, CosineKitty, Cryinghill, Curt Massie, Cvillar, Cygnis insignis, CyrilleDunant, D0762, DDima, DJ K666, DV8 2XL, DVD R W, Dali, Dancter, Danger, Danh10105, Daniel Case, Daniel J. Leivick, Dany680, Dar-Ape, Darklilac, Darkphotn, Darkwraith, David Eppstein, DavidLevinson, Davidstern, Dawn Bard, Dawnseeker2000, De Officiis, Deadkid dk, Deanex, Delete121, Dellexxx, Den fjättrade ankan, Denisarona, Denseatoms, Deor, Der Golem, DerHexer, Dethme0w, Diannaa, Dieselbub, Dietrichvonschnitzle, Direvus, Discospinster, Dispenser, Djcraize, Dmcq, Dmk5717, Dmprutland, Dnkidd, Dogears, Don't give an Ameriflag, DonSiano, Donarreiskoffer, Dontworry, Dougmcdonell, Doulos Christos, Dr Mullet, Dr.RonaldLett, Dreadstar, Drewlynn2138, Drmies, Dtobias, Ducttapeandzipties, Dullfig, E. Fokker, E2eamon, ESkog, Eaglizard, EdJogg, Eekerz, Efrankenberger, El C, Electron9, Element16, Elerium, Elert, ElinorD, Eljefe3126, Elkman, Emohareb, EncMstr, Engrocorp, Enric Naval, Enviroboy, Epbr123, Ephebi, ErasmusEros, Eric Kvaalen, Eric Winesett, Ericoides, Erikhansson1, Escape Orbit, Esowder, Etugam, Euchiasmus, Evpope, Excirial, ExtremeTube, Fancy steve, Farmercarlos, Fatality War, Felyza, Fieldday-sunday, FiggyBee, Flair Girls, Flux.books, FocalPoint, Fram, Frankschots, Fratter, Fritzpoll, Froid, Frymaster, Funandtrvl, Fwed66, Gaius Cornelius, Gary King, Gene Nygaard, GeoGreg, Getafix, Gfoley4, Ghewgill, Ghosts&empties, Giants27, Giftlite, Glacialfox, Glane23, Glenn, Glloq, Gmansport22, Godfollower4ever, Gogo Dodo, GoingBatty, GraemeL, Grafen, GreenReaper, Greg5030, GregorB, Gregorydavid, Grika, Grim23, Gun Powder Ma, Gurch, Gwernol, Gz33, H2ologged, HFtT4life, HKT, Hadal, Hagerman, Haikupoet, Hairy Dude, HalfShadow, Haljackey, HappyJake, Harryboyles, Hcheney, Hellbus, HelloAnnyong, Hellounique, Hemmer, HenryLi, Heracles31, Hericlius, Heron, HexaChord, Hithisishal, Hmains, Hmilgram, Hooba17, Hopingforgood, HornColumbia, Hottentot, Hr oskar, Hrco, Hwyengineer47, IceCreamAntisocial, Illcommunication, Im caius, Imgaril, Improver 03 04, Indon, Ingolfson, Intersofia, Intranetusa, Invierno14, Ionescuac, Iridescent, Ishokunaeppy, Isnow, J Milburn, J.delanoy, JAn Dudík, JMT, JaGa, Jacoblimlokguan, Jadewebster, Jagged 85, JamesBWatson, Jamescocks, Jamiefudge, Jayherrod007, Jazza18, Jeanpetr, Jeff G., Jeremy Visser, Jerry, JerryFriedman, Jglumb, Jhutch2, JidGom, Jimrickshaw, Jj137, Jkbice, Jm3, Jmrowland, Jnesbitt49, Joanne1689, JodyB, Joe Sewell, Joepirola, John, John Sheu, Johnpseudo, Jojhutton, Jon Harald Søby, Jonks, Jose77, Josh Parris, Joshua Issac, Jpc327, Jreferee, Jusdafax, KGasso, KVDP, Kasey mariko, Kate, KathrynLybarger, Kaykay246, Keraunos, Kernoz, Kevin kaior, Kinaro, King aardvark, Kingcnut, Kingpin13, Kipediwi, Kmac1036, KnowledgeOfSelf, Koavf, Kocisz, Kpeyn, Kuru, Kurykh, Kylu, LAX, Lacroselad, Lamro, Landon1980, Lateg, Lazarus1907, LeCaire, Leadplay, Leapman, Lemming64, Leszek Jańczuk, Levineps, Levskaya, Lightmouse, Lights, LindaKleine, Ling.Nut, LinguisticDemographer, Lockesdonkey, Lockley, Loren.wilton, LorenzoB, Lumber Jack second account, Luxgineer, MADe, MARussellPESE, MASQUERAID, MIRight, MK, MMS2013, Mac, Macy, Madassgamercal, Maid Goofs, Mainehcky45, Maksbul.tel, ManIsWhiteMonkeyIsBrown, Mangunikinkini, Manop, MarcMontoni, Marcod'Itri, Marechal Ney, Marek69, MariaElvira, Marshman, Martindelaware, Materialscientist, Mathonius, Mattisse, Mattjs, Maxí, Mejor Los Indios, Mephistophelian, Mets501, Mhia88, Michael Chusid, Michael Hardy, Michael Rawdon, Mihsa88, Mikeeg555, Mild Bill Hiccup, Miltonhowe, Mion, Miq, MkClark, Mkomplete, Mlhodges, Mm40, Mmele, Mmm, Monty845, Mr Stephen, Mr. Carpenter, Mr. Nacho de la Libre, MrKoch1900, Mrqwerty987, Mryakima, Mspurlock73, Murugasethu, Mwanner, Mwtoews, N1RK4UDSK714, NHRHS2010, Nabokov, Naohiro19, Narayansg, Natalya, NawlinWiki, Neelix, NellieBly, Neo139, Nepenthes, Nerdo18, Nevsal, Newstateofme, Nhoman, Nick Number, Nk, Nkayesmith, Nonagonal Spider, NortyNort, Nuujinn, Nvrsmmr, Octane, Oda Mari, Ohconfucius, Oldfag1029384756, Onore Baka Sama, Orangemike, Orangutan, Oscarthecat, OverlordQ, Oyvind, PRiis, Parker007, Patche99z, Patstuart, Paul August, Pauli133, Peter Delmonte, Peter Horn, Peter.C, Peterlewis, Petrb, Petri Krohn, Pgk, Phantomsteve, Phasmatisnox, Philbert2.71828, Philip Trueman, Piano non troppo, Pigsonthewing, Pinethicket, Pingveno, Pirkid, Pne, Poetryrus, Pol430, Pollinator, Polly, Polyparadigm, Polytrope, Ponyo, Prenigmamann, PrestonH, Pro crast in a tor, Proyster, Puckly, Pushcreativity, Quadell, QuadrivialMind, QuantumEleven, Quintote, Qwert, RHaworth, RJaguar3, RL0919, Radio Flyer Guy, Rainbowflowerchild, Rallas202, Rama, Randomer 102, RapidSet, Raymondwinn, Razorflame, Realitybend, Reaper Eternal, Rebrane, Redsully, Redvers, ReedConstruction, Reedy, Requestion, ResearchRave, RexNL, Rfquerin, Rfsonders, Rhobite, Rhopkins8, Rich Farmbrough, Rich257, Richwhite10, Rjwilmsi, Robertbian, Robfwoods, Romanm, Ronhjones, Ronz, RoyBoy, Royalbroil, Rrodden, Rror, Rustalot42684, Rvsingh83, Ryulong, Saejinn, Saintrain, Samohyl Jan, Sardanaphalus, Savolya, Schmidt455, Schrodinger's cat is alive, SchuminWeb, SecretDisc, SeventyThree, Shaddack, Shanes, Shinkolobwe, Shoaler, Shushruth, Signalhead, Siim, Siltsalt, SimonScowen, Sir Vicious, Sjanes71, Skiffer, Skifree11, Skunkboy74, Smack, Smalljim, Smbrown123, Smcgough, Snake doctor65, Snottily, Solsikche, Some jerk on the Internet, Some thing, Sonett72, SpaceFlight89, Spangineer, Speedevil, SpikeToronto, Spiralwind, Srkarade, Steinsky, Stellacad, Stephenb, Steven Walling, SteveofCaley, Stockful Guy, Stormwriter, Struthious Bandersnatch, Sumsum2010, Sunyoh, Supenerd, SusanLesch, T, TFNorman, TVBZ28, TYelliot, Tahavur, Tangotango, Taxman, Tbhotch, Tcncv, Tedd, Terencehill, Thatguyflint, The Thing That Should Not Be, The undertow, TheProject, TheRingess, Thegodspeeder, Thehelpfulone, Thewoodsman, Thingg, Thinkjones, Thomasmc14, Thondulkar, Thrind, Thrissel, Thumperward, Tide rolls, Timwi, Tingrin87, Titoxd, Tivona schneider, Tkn20, To satyajit73, Tobby72, Toiyabe, Tom harrison, Tomboy Man, Tommaisey, Toni baptista, Towerman86, Traal, Tregoweth, Treisijs, Trevor MacInnis, TreyGeek, Triona, Triskele Jim, Troy Frei, Trusilver, Trythisonyourpiano, Tucjer, TundraGreen, Turbo Dragon, UltraMagnus, Uncle G, Unint, UnitedStatesian, Unyoyega, UpstateNYer, Utcursch, VMS Mosaic, Van helsing, Vanished User 1004, Vanished user 39948282, Vegaswikian, Veinor, Velella, Vfridman68, Violetriga, VirtualSteve, WJBscribe, WTF Formwork, Walkerma, Warut, Washburnmav, Wavelength, Wee-Kee-Pee-Dee-Ah, Wellmer, Wendyfables, WereSpielChequers, WhisperToMe, Whomp, Why Not A Duck, WikiHead, Wiki alf, WikiSSI, WikiTome, Wikitumnus, Will Beback Auto, Willp4139, WilyD, Wimt, Witan, Wolframfranke77, Wsiegmund, Wtshymanski, Xanzzibar, Xb-70, Yaris678, Yath, Ysharlat, Zarius, Ze miguel, Zebra07 uk, Zeimusu, Zhar, Zian, Zig573, Ziplock7, Zuejay, Zzuuzz, ~shuri, Δ, 1452 anonymous edits

Portland cement *Source*: http://en.wikipedia.org/w/index.php?title=Portland_cement *Contributors*: AHEMSLTD, Aboutmovies, Ae-a, Alberto Cyone, Alchemist2001, Aldie, Ale jrb, Ali@gwc.org.uk, Alma Pater, Argyriou, Armbrust, Astronautics, Attilios, AxelBoldt, BD2412, BRG, Ballista, Bastin, Bbeggs, Before My Ken, BillyPreset, BlueCanoe, Blugill, Bobblewik, BoulderDuck, Brambleclawx, Bryan Derksen, Bushytails, Buster7, CFB123, ChemGardener, Chesscubes, Chris 73, Chris the speller, Chzz, Cimento.Org, Click23, Colinsweet, Cryinghill, Crzrussian, CyminX, CyrilleDunant, D6, DMChatterton, DanMS, Darklilac, Dave.Dunford, Detayls, Dlabtot, DogFoxen, Dogcow, DonSiano, Donarreiskoffer, Dougmcdonell, Dwhargreaves, EPO, EdBever, Edison, Edward, Element16, EoGuy, Excirial, Fixmacs, FocalPoint, Foxandpotatoes, France3470, Gene Nygaard, GeoGreg, Georg Locher, Gierczyk, GoatGuy, Goffrie, GoingBatty, Gregorydavid, Hemmingsen, Hephaestos, Heron, Hooperbloob, Ilovewp, Iridescent, J.delanoy, JJ Harrison, JJackson8, JSpung, Jeff G., Jeni, Jhog1978, Joetheguy, Jojhutton, Jordan Elder, Jpbowen, Julesd, KDS4444, KVDP, Karl-Henner, Kirachinmoku, Kpeyn, Lexowgrant, LinguisticDemographer, MARussellPESE, MER-C, MIckStephenson, Mac, Mak17f, Mash morgan, Materialscientist, MauriceJFox3, Mcginnly, Meisam, Mfwitten, Michael Hardy, MichaelJanich, Milton Cintra, Mjk2357, Moormand, Mvc, Nabokov, Naufana, NawlinWiki, Nikai, Niro5, Nwbeeson, Oxymoron83, Oyvind, PAlc, Pacific1982, Pauli133, Paypaypaymean, Pbenken, Pdcook, Pfidjes, Phantomsteve, Pinethicket, Plugwash, PoccilScript, Postrach, RJP, Radon210, Remi0o, Rhallanger, Rhanyeia, Rhebus, Rich Farmbrough, Richard Catto, Richard New Forest, Rjwilmsi, Rohit.k88, Russ London, Shinkolobwe, Shrew, SimonScowen, Sloppy, Snigbrook, Sonett72, Spangineer, Special-T, Steinsky, SusanLesch, TDogg310, Tabletop, Terminator484, Thingg, Tim!, Tom Permutt, Toni baptista, Topken Microsilica, Trusilver, Twunchy, Ultraseann, VMS Mosaic, Van helsing, Vsmith, WAS 4.250, Wes Hermann, WikiHead, Wimt, Wknight94, Wleizero, Woohookitty, Wsiegmund, Xandi, Xanzzibar, Yyy, Ziggy Sawdust, ZorPrime, 319 anonymous edits

Tailored Fiber Placement *Source*: http://en.wikipedia.org/w/index.php?title=Tailored_Fiber_Placement *Contributors*: Erfurth, Mlaffs, R'n'B, SPI IPF, Tony1, 4 anonymous edits

Composite material *Source*: http://en.wikipedia.org/w/index.php?title=Composite_material *Contributors*: 12 Noon, 777sms, 956391, Ahering@cogeco.ca, Ahoerstemeier, Aitias, Akkida, Ale jrb, Alexius08, Amgreen, Andrewpmk, Andries, Ankurtg, Anna Lincoln, Apollionus, Archer3, AshishG, Aushulz, AxelBoldt, AzaToth, BD2412, Bardsandwarriors, Barek, Beetstra, Bejnar, BenBreen2003, BenFrantzDale, Bertiethecat, BilCat, Bkumartvm, Bluemoose, Boubacar, Brookie, Bryan Derksen, C+C, Cases2Go, Cb2292, Cchambers, Cherry.charan, Chris the speller, ChyranandChloe, Closedmouth, Cmprince, Compositedoorshop, Compositeguy, Contaldo80, Coosbane, Coxt001, Crazyhug, Creasyts, Cstaffa, Cureden, Dan100, Darkwind, December21st2012Freak, Deltasquared, Denzil Simoes, Discospinster, DivineAlpha, Dolphin51, Donbert, Dr eng x, Dsajga, Durkee, Dwayne, Ecs4life1, Elasint, Eliashedberg, Elliskid, Ellywa, Enviroboy, Eoinsimon, Epbr123, Facts707, Falcon8765, Farokh mehr, Femto, Fieldday-sunday, Flippythewalrus, Fnlayson, Foxcreekcowboy, Frap, G. Hill, Gardar Rurak, Geoffrey Wickham, Gilliam, Golgofrinchian, Gosgood, Greg L, Greudin, Gzuckier, Hadoooookin, Haftchen, Hans Dunkelberg, Henning Makholm, Herbythyme, Heron, Hu12, Hydrargyrum, Hydrogen Iodide, Ian Pitchford, Iediteverything, Igoldste, Incompetence, Indon, Inwind, JTSchreiber, Jag123, Jain034567, Jakerome, Jamusmax, Jandrel, JayjayVicious, Jeff G., JidGom, Jimaginator, Jmrowland, Jnavas2, John, Jomasecu, Joostvandeputte, Jpo, Jsummerscales, Jthoele2, Jusdafax, Kandi111777, Keegan, Kerina yin, Ketiltrout, Kingpin13, Kipediwi, Kirt Butler, Kku, Kokcharov, Kovianyo, Kuru, Kuzaar, Kvdveer, KyraVixen, LADave, Ldemasi, LeaveSleaves, Leumar01, LiDaobing, Lights, Lotje, Luigizanasi, Luwilt, MER-C, Mairi, Makemi, Marcus Qwertyus, Martijnrd, Materialscientist, Mbvanleeuwen, Mbw5014, Mecanismo, Michael Hardy, Michaelbusch, MidgleyDJ, Miladmilad, Mion, Mmeijeri, Moeron, Mvelterop, Naohiro19, Nasier Alcofribas, Natalya, Nathaniel, Nbarth, Nbvvbn, Netcomposites, Nono64, NorwegianBlue, Nuno Tavares, Oberst, Occamisation, Oleg Alexandrov, Oli Filth, Onionmon, OverlordQ, Owen, Patelurology2, Paulbracegirdle, Peterlewis, Pgk, Philander, Physchem, Plamal, PlatinumX, Polyparadigm, Possum, Prasun92, Pritam79, Pyrope, Qwertythecat, RHaworth, Ranjithsutari, RedWolf, Reedy, Resuna, Rich257, Robert Mathel, Romaioi, Ronhjones, Rror, RuM, SME2009, SMcCandlish, SPI IPF, Sam Korn, Scohoust, Scottanon, ScottyBerg, Sennaya, Shaddack, SimonP, Sivakamitvm, SkyWalker, Slakr, Slashme, SnappingTurtle, Sole Soul, SouthLake, SpaceFlight89, Spitfire19, Sprotopapas, Stardust8212, Stephenb, Stephen Leonard-Williams, Steve2011, Student1980, Suisui, SummonerMarc, Surendracomposites, Sylvain Mielot, TShilo12, Tanvir Ahmmed, TastyPoutine, Teapeat, TekWrtr, The Thing That Should Not Be, Thingg, Think outside the box, Tlesher, Tomcox, Tyco.skinner, Ultramandk, UncleDouggie, Useight, VK35, Vadim Makarov, Vaidheestvm, Vegaswikian1, Veinor, Versageek, Vikramaditya1986, Vtaber02, Webmasters, West.andrew.g, WikiHead, Wizard191, Yosri, Zfr, Ásgeir IV., , 572 anonymous edits

3D composites *Source*: http://en.wikipedia.org/w/index.php?title=3D_composites *Contributors*: Cesium 133, Fabrictramp, Katharineamy, Malcolma, Yutsi

3T Cycling *Source*: http://en.wikipedia.org/w/index.php?title=3T_Cycling *Contributors*: Harman mogul, Headbomb, JaGa, M-le-mot-dit, Midway, PhnomPencil, Postcard Cathy, R'n'B, Severo

Fiber cement siding *Source*: http://en.wikipedia.org/w/index.php?title=Fiber_cement_siding *Contributors*: Abductive, Acroterion, Alan bron, Angela elliss, ArielGlenn, Barek, Billbeee, Boby-27, Bryanclair, Cadwallader, Cafemaster, Calton, Clippernolan, Hamiltonstone, Holger.Ellgaard, Hooperbloob, Hu12, I82much, Insomnivore, Inwind, Jcfretts, Jmundo, Jon ochshorn, Jorge1000xl, Kevinkor2, Kkmurray, Kymacpherson, LedgendGamer, Liftarn, Macrakis, Mahanga, Masterfulmind2007, Mccawii, Melnick25, Ntsimp, Old Moonraker, Pengo, Pjrosch, Shag1116, Sidingpro, Synstone, TVBZ28, Tangerineduel, Twas Now, Warut, Zeamays, 54 anonymous edits

Fiber pull-out *Source*: http://en.wikipedia.org/w/index.php?title=Fiber_pull-out *Contributors*: Kerina yin, Leumar01, Malcolma

Fiber-reinforced composite *Source*: http://en.wikipedia.org/w/index.php?title=Fiber-reinforced_composite *Contributors*: Alan Liefting, Andy Dingley, CommonsDelinker, Ecodek, Frap, Kerina yin, Look2See1, Materialscientist, Pingfan, Quintus fabius, SVTCobra, Some standardized rigour, 29 anonymous edits

Nu-Wood Decorative Millwork *Source*: http://en.wikipedia.org/w/index.php?title=Nu-Wood_Decorative_Millwork *Contributors*: Belovedfreak, Blainster, Chercod1, Cloveious, DragonflySixtyseven, Edward, Gilliam, Heron, Islandboy99, Jdcooper, Josh Parris, JustSomeKid, Kevinkor2, LinguistAtLarge, Malcolma, Notthe9, Pburka, PierreAbbat, Robin Patterson, Satori Son, 7 anonymous edits

Fiber volume ratio *Source*: http://en.wikipedia.org/w/index.php?title=Fiber_volume_ratio *Contributors*: D6, Fabrictramp, Kerina yin, Malcolma, Markiewp, Tassedethe

Fiber-reinforced concrete *Source*: http://en.wikipedia.org/w/index.php?title=Fiber-reinforced_concrete *Contributors*: -Majestic-, 12 Noon, 7, Afluegel, Alcuin, BillFlis, Blabbermouth01, Boinkman27, DMChatterton, Dave seer, Falcon8765, Farokh mehr, Feureau, Frap, GFRC, Geoing, Goldenrowley, IMM39, Jmlazarus, Kowalma, Leranedo, LilHelpa, Makemi, Mnyaseen, MoogleDan, Moreschi, Oculus Tauri, Paleorthid, Paulbracegirdle, Pearle, RJFJR, Rrburke, Some standardized rigour, Srobak, Svick, The Captain Returns, Thumperward, Topbanana, Welsh, 46 anonymous edits

Orangeburg pipe *Source*: http://en.wikipedia.org/w/index.php?title=Orangeburg_pipe *Contributors*: Anonymous Dissident, Astrovega, Dan Sarandon, Hydrargyrum, Luvcraft, Vengeful Cynic, Zondi, 5 anonymous edits

Image Sources, Licenses and Contributors

Printed by Books on Demand GmbH, Norderstedt / Germany